How to Make Planes, and Vices

Written and illustrated by Aaron Moore

Step-by-step instructions on how to make: JACK PLANES, REBATE PLANES, PLOUGH PLANES, SPOKESHAVES, SASH CRAMPS, BENCH CRAMPS and VICES.

Intermediate Technology Publications 1987

Acknowledgements

Financial assistance in the production of this manual was made available through the Intermediate Technology Development Group, from a grant given by the Overseas Development Administration. Their assistance is gratefully acknowledged.

© Intermediate Technology Publications 1987
9 King Street, London WC2E 8HW, U.K.

ISBN 0 946688 98 2

Printed by the Russell Press Ltd., Bertrand Russell House,
Gamble Street, Nottingham NG7 4ET, U.K.

Contents

	page
Introduction	1
The tools	2
Glossary	3
Jack plane	5
— handled version	21
Rebate plane	33
— adjustable fence	44
— alternative fence assembly	52
Plough plane	55
Spokeshave	69
Sash cramp	79
Bench cramp	87
Leg vice	99

ERRATA
p.35 — Part D: The dimensions should read 250 × 65 × 20
p.57 — Part B: The dimensions should read 250 × 35 × 25
p.57 — Part E: The dimensions should read 250 × 50 × 25

Introduction

This manual describes in detail how to make seven different woodworking tools. Directions for their construction and use are in the form of step-by-step illustrations, backed up by short descriptive texts. Most of the information is supplied by the drawings, but it is important to read the captions carefully, because it is impossible to draw every detail needed for construction.

The tools described here are all very practical and cheap to make and include a JACK PLANE, REBATE PLANE, PLOUGH PLANE, SPOKESHAVE, SASH CRAMP, BENCH CRAMP and a type of BENCH VICE. All the planes except the jack plane are quite specialized tools, but they are essential pieces of equipment for a rural workshop, with no electricity, if it is to produce good quality work, efficiently. The cramps and vices are also essential. Without them, work cannot be cramped and glued properly, or held firmly enough for cutting joints accurately.

These tools have been developed to be made in situations where money is simply not available to equip a workshop with expensive, imported western tools. They are appropriate for both large training institutions, where the students can make tools for the school and for their own use, or for small village workshops where the craftsman can make his own tools as and when he needs them. It may even be possible to set up small tool-making businesses, supplying schools, colleges and shops in the surrounding area.

This is not a carpentry text book: I have assumed that the reader has a basic knowledge of woodwork, that he is capable of preparing timber to size, that he is familiar with a number of simple woodworking techniques, and has the enthusiasm to overcome setbacks and mistakes.

To begin with, a bench and a good kit of tools will be needed. This would include: TRY SQUARE, MORTISE GAUGE, JACK PLANE, HAMMER, CENTRE PUNCH, HALF-ROUND RASP, MALLET, VARIOUS CHISELS, SCREW DRIVER, WHEEL BRACE, CARPENTER'S BRACE, a SET OF BITS and an OIL STONE. Not all of these tools are needed to make each design. It is up to the reader to decide which tools are made, bearing in mind the materials and equipment available, and the requirements of the workshop.

The quality of the tools described in this manual depend a great deal on the workmanship and materials used. In many cases the metal parts will have to be bought, but whenever possible use the best timbers, and take as much time and care in construction as you can.

None of these tools are perfect; they may require practise to use properly, they may even break — but compare the cost of a home-made tool and the cost of a similar tool in a shop. It may be up to one-third cheaper. Also consider the problem of a broken shop-bought tool: spare parts are expensive and often unobtainable in a developing country, and the cost of replacement will be greater than the original cost due to inflation. To repair a home-made tool may cost next to nothing. Of course there is nothing to stop you buying tools once your workshop is making money; on the other hand you may find it unnecessary. But in the beginning, is there any other way of starting out with little or no support?

1

The tools

Jack plane

Wooden planes take time and skill both to make and to use, but once the skills are mastered these tools can give a great deal of satisfaction. Treated with care and respect a wooden plane made from good quality hardwood will work as well as any metal plane. Eventually the sole will become worn, but it is a simple matter to true it up again with another plane. When the mouth becomes too large, a small piece of wood can be inserted in the front to fill the gap.

Plough and rebate planes

Of the two, the plough plane is the most difficult to make, because of the metal sole plate. However if it cannot be made by a woodworker, it is definitely not beyond the capabilities of a rural metalworker.

Both of these tools have been designed so that a chisel can be used as the cutting edge, which means that even if one or two chisels do have to be bought, they will have a dual function in the workshop.

Spokeshave

The spokeshave is a very versatile tool, and when used in conjunction with a bow saw or a coping saw can produce decorative and pleasing products.

As with the plough and rebate planes, it is not necessary to buy a blade specially for this tool; a normal jack plane blade will do the job adequately.

Cramps and vices

All of the cramping tools described in this manual are made entirely of timber and use wooden wedges to apply pressure to the work piece. Take time to learn how to use these tools effectively without abusing the wedges with heavy mallet blows.

The leg vice, described on page 101, could be modified quite easily into a screw vice, using a length of threaded steel rod and a tommy bar, but this would be expensive.

The dimensions in this manual are all in millimetres, and for best results they should be adhered to quite strictly unless otherwise stated. Before making a tool, read through the text and follow the drawings until every detail is understood. In many cases the sequence of work is important.

Choose the timber to be used carefully. It must be hard, with close straight grains, no knots or splits, and it must be dry. The wooden parts of these tools can be finished with sandpaper and coated with linseed oil or varnish. Metal parts should have any sharp edges smoothed off with a file, and could be painted with enamel paint.

Hopefully the manual will stimulate ideas and imagination, and the reader will think twice before walking into a tool shop.

2

Glossary

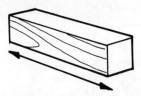

Grain
The grain is the lines and patterns seen on the surface of a smooth piece of wood. The arrows show the direction of the grain.

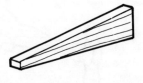

Wedge
A wedge is a piece of timber with its edges forming a shallow point at one end.

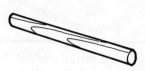

Dowel
This is a small pin of wood with a round cross-section. It is often used to fix wood joints together instead of nails.

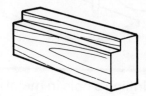

Rebate
A rebate is a rectangular recess or step along the edge of a piece of wood.

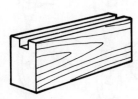

Groove
A groove is a channel or a hollow cut into one side of a piece of wood.

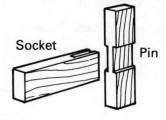

Bridal joint
The two parts of this joint consist of a socket and a pin.

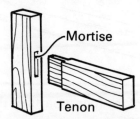

Mortise and Tenon joint
This consists of two parts, the mortise, which is a square or rectangular hole, and the tenon, which fits securely into the mortise.

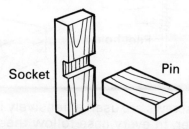

Housing joint
The two parts of this joint consist of a socket and a pin.

DRILLING METAL

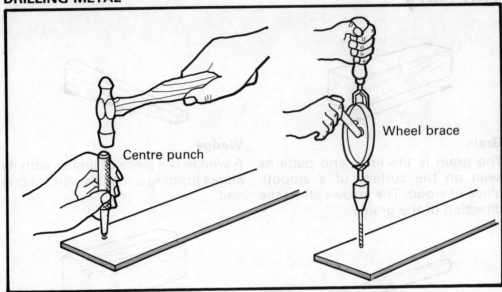

When drilling a hole in metal it is important to punch a small dent in the work piece to prevent the drill bit from wandering off the mark.

FIXING TIMBER WITH WOODSCREWS

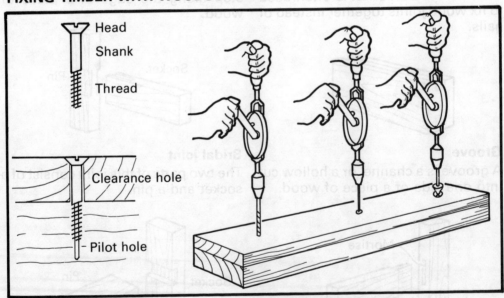

Woodscrews are used extensively in this manual for fixing parts of the tools together. In every case follow these instructions:
1. Drill a clearance hole through the top component, big enough to allow the shank of the screw to pass through easily.
2. Drill a pilot hole into the bottom piece. This should be small enough to give the thread of the screw a good 'grip'.
3. Countersink the top component for the screw head.

Jack plane

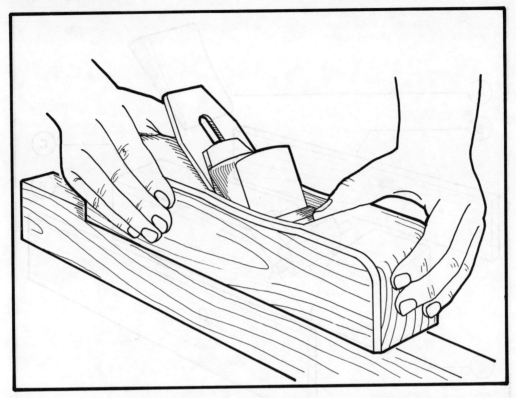

The jack plane is the most useful size of plane for general carpentry and joinery work. It can be used for preparing timber to size, trueing up boards for edge to edge joints, and for smoothing down jointed work after it has been glued together.

The dimensions given here can be altered quite easily to produce more specialized tools. To make a TRY PLANE, prepare the timber for the stock to about 500mm in length. For a SMOOTHING PLANE make the stock 200 or 250mm long. When altering the dimensions remember that the mouth opening should be 1/3 of the length of the plane back from the front.

Before you begin on this project it is important that a good plane blade and back iron is obtained, because this will determine the width of the stock. Also decide which design to follow. The first is a simple box-shaped plane, the second, a more complicated version with knob and handle.

The two most important aspects of plane making are to make the mouth opening the right size, and the throat big enough to allow the shavings to pass through easily.

Once the techniques for making this tool have been mastered it is a simple step to move on to more specialized moulding planes or low-angle block planes.

LIST OF PARTS

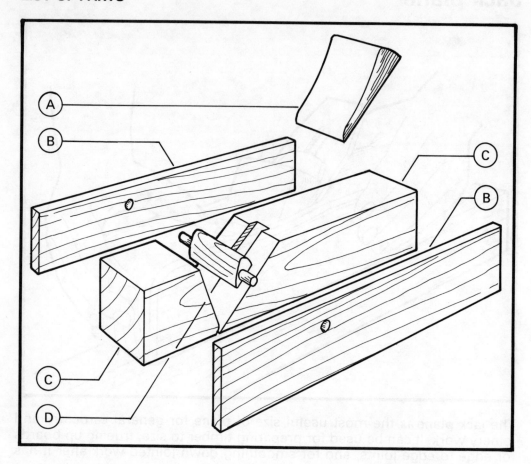

Part

A *The wooden wedge* is shaped to fit between the wooden cross bar and the blade, which it locks tightly in position.

B *The side pieces* are glued to the central stock, and form part of the body of the plane.

C *The stock* is made from one piece of timber, cut and shaped to form the throat and the bed for the blade.

D *Wooden cross bar*. This is fitted into two holes in the side pieces. To make a really tight fit for the wedge it is essential it is allowed to move freely, so it must not be glued.

CUTTING AND PARTS LIST

Part	Name	Quantity, Material and Dimensions (mm)
A	Wedge	1 pc. Timber 150 × 25 × 'Dim A' + 34
B	Side pieces	2 pcs. Timber 350 × 75 × 15
C	Stock	1 pc. Timber 350 × 75 × 'Dim A' + 4
D	Wooden cross bar	1 pc. Timber 'Dim A' + 34 × 30 × 20

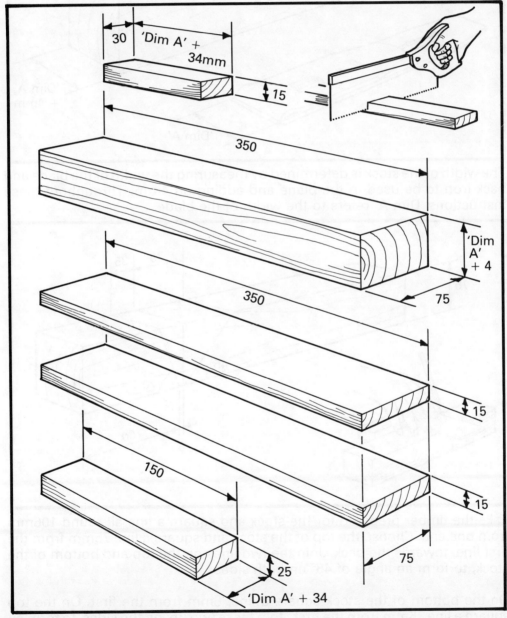

Diagram of cutting list.
'Dim A' refers to the width of the blade.

MAKING THE STOCK

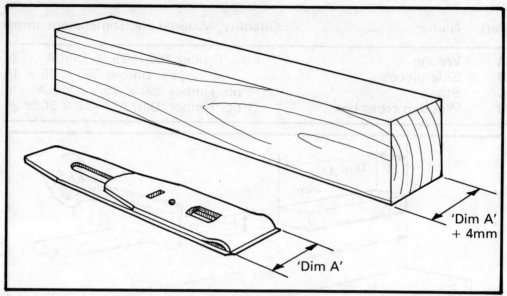

The width of the stock is determined by measuring the width of the blade and back iron to be used in the plane and adding on 4mm. Throughout these instructions 'Dim A' refers to the width of the blade.

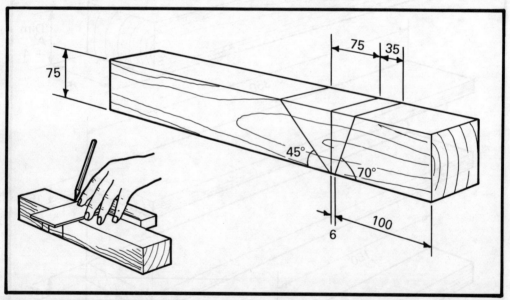

Take the timber prepared for the stock and square a line all round 106mm from one end. Choose the top of the stock and square a line 75mm from the first line, towards the back. Join the two lines on the top and bottom of the stock, to form an angle of 45° on both sides.

On the bottom of the stock square a line, 6mm from the first. On the top square a line 35mm from the first. Join these two up, on the sides, to form an angle of about 70°.

8

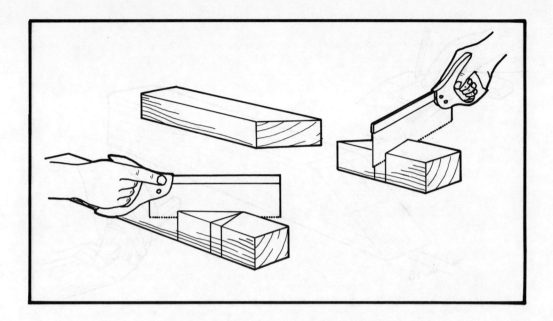

Cut the stock into three pieces. The centre section is waste, so take care to cut on that side of the line.

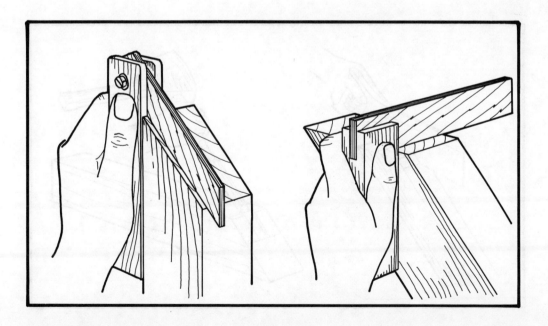

Take the largest section of the stock and plane the angled end, making sure it is square to the sides and at an angle of 45° to the bottom.

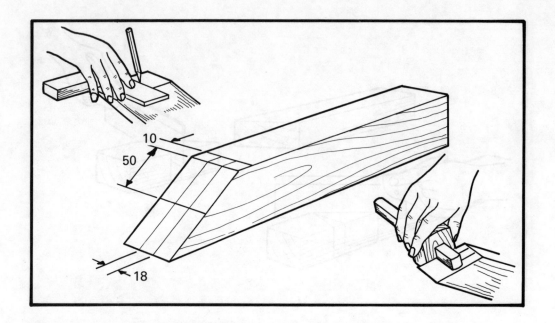

To mark the recess for the back iron screw, gauge two lines down the centre of the angled end, 18mm apart.

On the top of the stock square a line 10mm from the edge. On the angled end, square a line 50mm down from the edge.

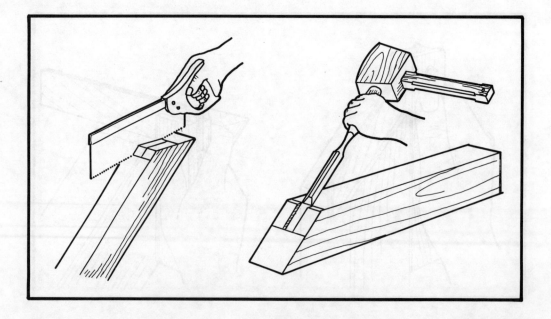

Saw the sides of the recess with a tenon saw and cut out the waste with a mallet and chisel.

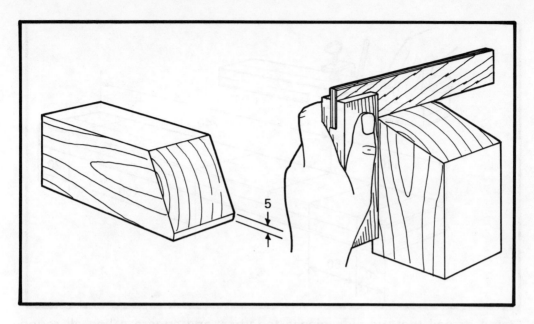

Take the smaller section of the stock and use a plane to square up the angled end, also round off the sharp edge slightly. This will prevent the mouth opening getting too big when the sole is trued.

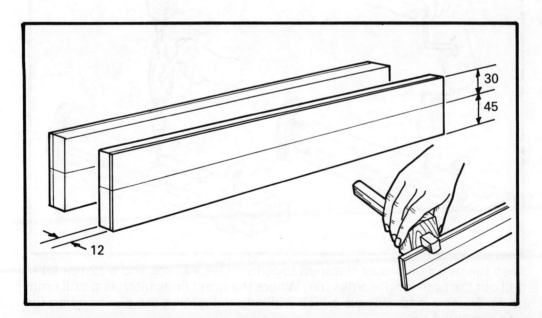

Take the two side pieces and gauge their thickness to 12mm. The waste will be planed off after the plane has been glued together. Also gauge two lines 45mm from the edge of both pieces.

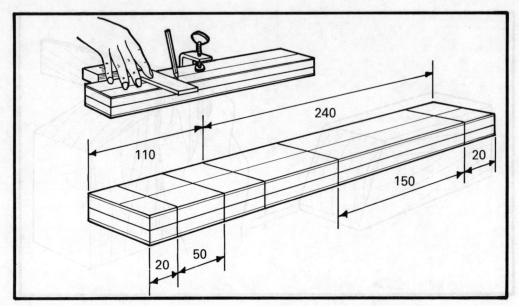

Clamp or nail the two side pieces together, and square a line all round 110mm from one end. Where this line meets the gauge line the hole for the wooden cross bar will be drilled.

Square the other lines all round as shown in the drawing.

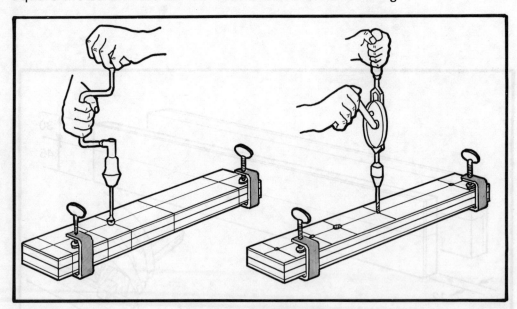

Keep the two side pieces clamped together. Use a brace and a 12mm bit to drill out the hole for the cross bar. Where the other lines intersect, drill 6mm holes. These are for dowels which will help in locating the pieces when the plane is glued up.

Always drill halfway through the workpiece, and turn it over to complete the hole.

MAKING THE WOODEN CROSS BAR

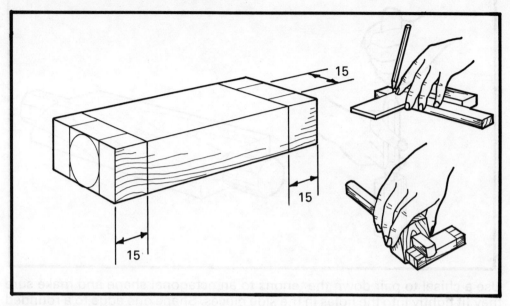

Take the timber prepared for the cross bar and square two lines all round 15mm from both ends.

Use a mortise gauge to mark 15mm tenons in the centre of both ends.

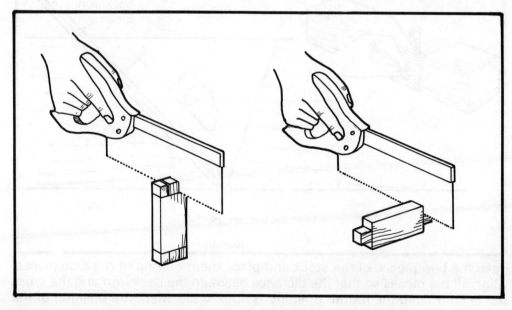

Use a tenon saw to cut the tenons on each end of the cross bar.

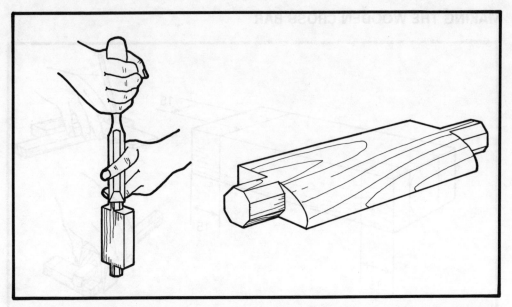

Use a chisel to pair down the tenons to an octagonal shape and make sure they fit tightly into the holes in the side pieces. Shape one edge to a rounded section to allow the shavings easy access out of the plane.

ASSEMBLING THE STOCK

You will need at least five clamps to assemble the plane. If there are not enough available, there are instructions on page 20 for screwing the stock together.

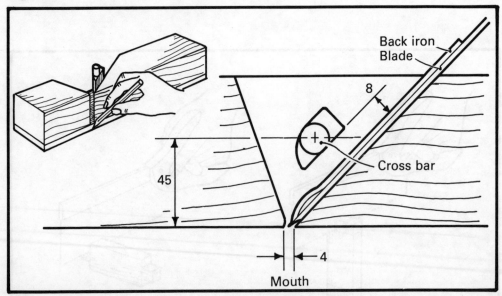

Take the two pieces of the stock and place them on one of the side pieces. Align all the pieces so that the distance between the back iron and the cross bar is 8mm, and the mouth opening is 4mm wide. Mark the position of the mouth on both side pieces.

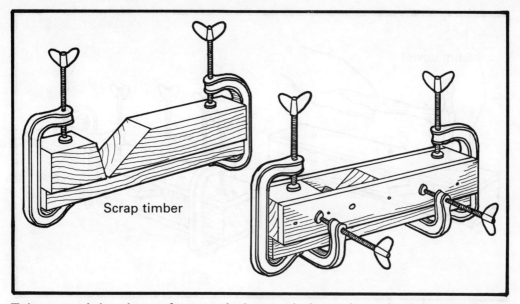

Scrap timber

Take a straight piece of scrap timber and clamp it to the bottom of both pieces of the stock, leaving a gap of 4mm between the two parts of the stock for the mouth.

Insert the wooden cross bar into the two side pieces and clamp these onto the stock. Make sure everything is aligned properly and the cross bar is parallel to the mouth opening.

Prepare eight 6mm dowels 25mm long.

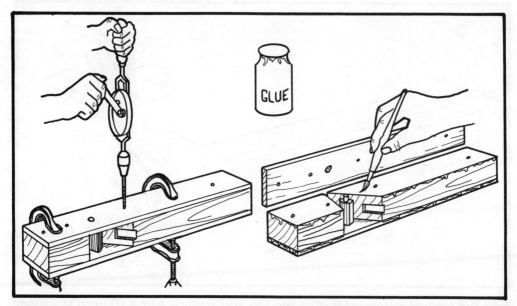

Use a wheel brace and a 6mm bit to drill into the stock through the eight holes in the side pieces.

Apply a good quantity of glue to the central parts of the stock only and assemble the pieces. **Do not forget the cross bar.**

15

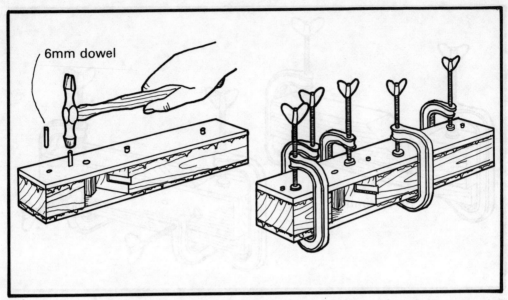

Drive the dowels into the stock and clamp the assembly together with as many clamps as possible. Leave the glue for at least six hours.

MAKING THE WEDGE

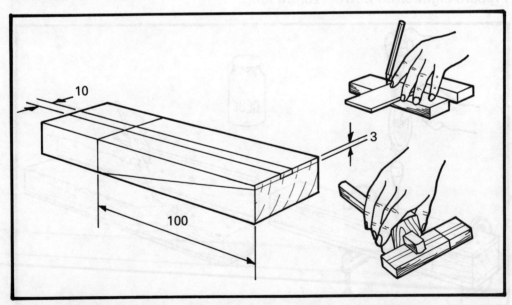

Take the timber prepared for the wedge and square a line 100mm from one end. Mark the shape of the wedge by drawing a diagonal on both edges.

Mark the groove for the back iron screw with two lines 10mm apart on one side only.

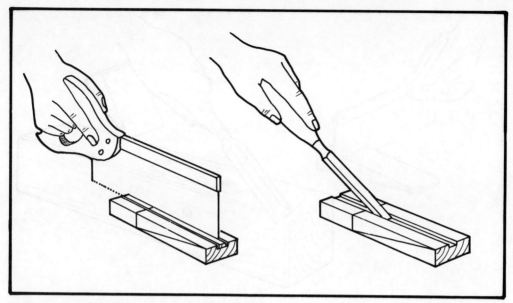

There are three ways to cut the groove:
1. Use a plough plane if one is available.
2. Use a cutting gauge, and remove the waste with a chisel.
3. Use a tenon saw to cut the sides of the groove and chisel out the waste.

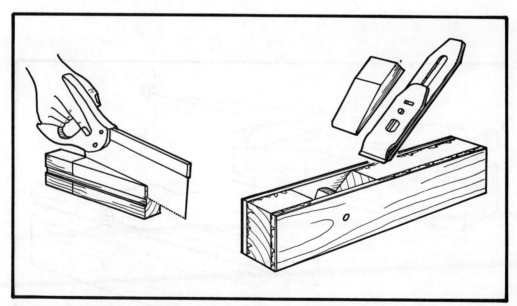

Saw the wedge to shape.

Take the blade, back iron and wedge and fit them into the throat of the plane. Take time to make the wedge fit tightly. It will need to be planed smooth, and may also need rounding off on both sides of the sharp edge.

17

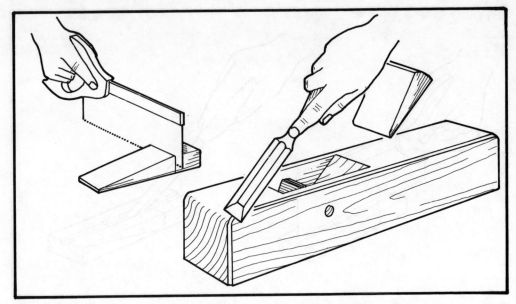

Saw off the top end of the wedge so it is about 100mm long.

At this point the reader must decide whether to carry on and make the plane on page 21, or stick to the simpler version. In the latter case the stock must be planed smooth and all the edges rounded off to make it comfortable to hold.

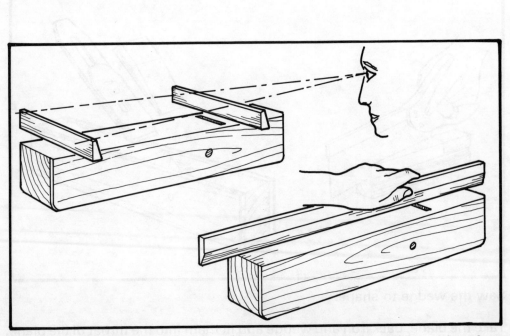

Plane the sole of the plane very carefully so that it is straight and free from winding. The mouth opening should be no wider than 5mm.

SETTING THE PLANE

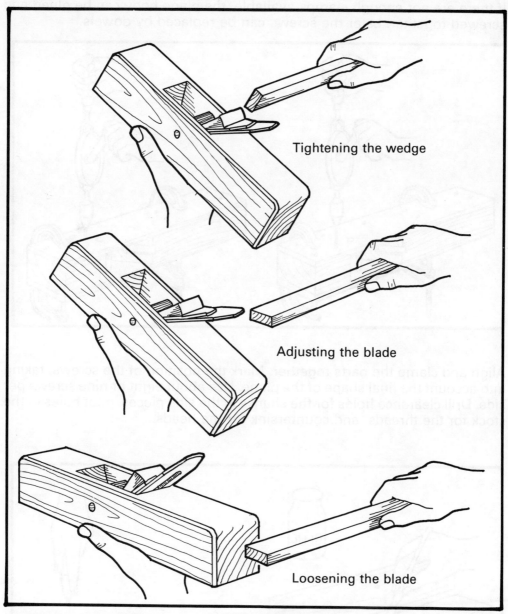

To tighten the wedge and the blade, strike the wedge with a piece of wood or a mallet.

To adjust the blade forward, strike the back of the cutter.

By hitting the back of the plane, you loosen the wedge and also bring the blade back.

ALTERNATIVE METHOD OF ASSEMBLY

If there are not enough clamps available, the plane body can be glued and screwed together. Later the screws can be replaced by dowels.

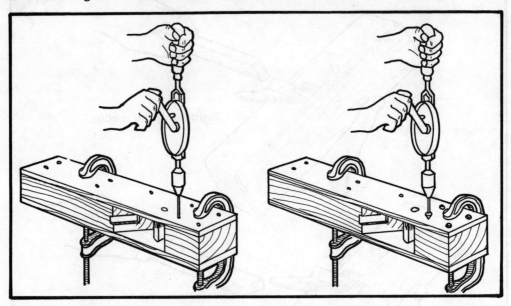

Align and clamp the parts together. Mark the position of the screws, taking into account the final shape of the plane. Use about eight or nine screws per side. Drill clearance holes for the shanks in the side pieces, pilot holes in the stock for the threads, and countersink for the heads.

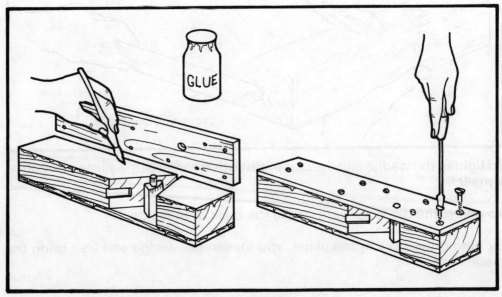

Glue and screw the pieces tightly together.

Jack plane, handled version

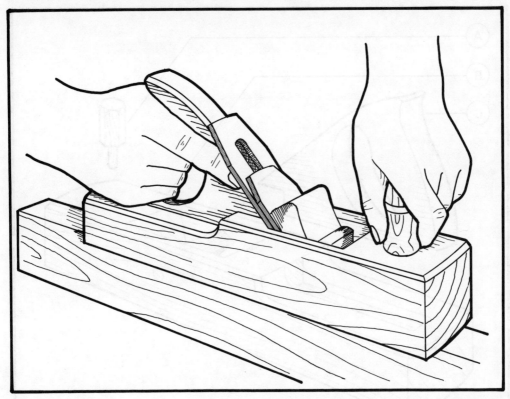

On the following pages are instructions for adding a handle and a knob to the stock previously described. Apart from making the plane more comfortable to hold and easier to control, these modifications can make the tool into a very pleasing object.

Both the stock and the handle need to be shaped and smoothed very carefully. A sharp chisel or knife, a rasp and sandpaper will be needed to do the job well.

LIST OF PARTS

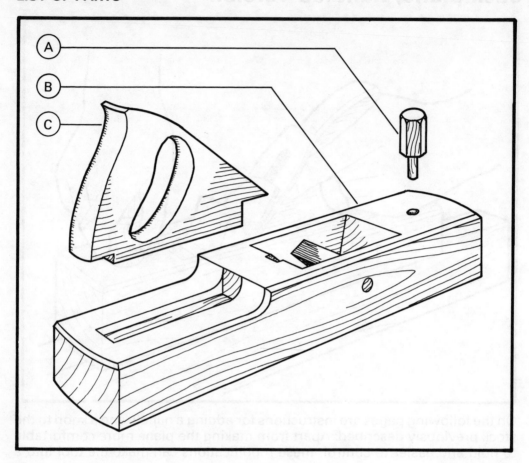

Part

A *The knob* can be turned on a lathe if one is available, or shaped by hand. It must be securely jointed to the stock.

B *The stock* is made by following the instructions beginning on page 10.

C *The handle*. This is cut and shaped by hand, and glued into a mortise cut out of the stock.

CUTTING AND PARTS LIST

Part	Name	Quantity, Material and Dimensions (mm)
A	Knob	1 pc. Timber 80 × 30 × 30
B	Stock	See pages 7-22 for details
C	Handle	1 pc. Timber 180 × 100 × 25

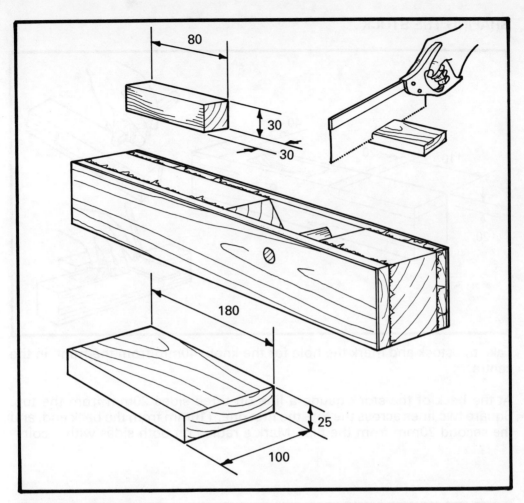

Diagram of cutting list.

SHAPING THE STOCK

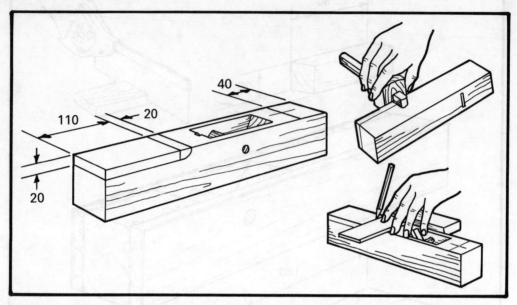

Take the stock and mark the hole for the knob, 40mm from the front, in the centre.

At the back of the stock gauge a line on three sides 20mm from the top. Square two lines across the top, the first one, 110mm from the back end, and the second 20mm from the first. Mark a radius on both sides with a coin.

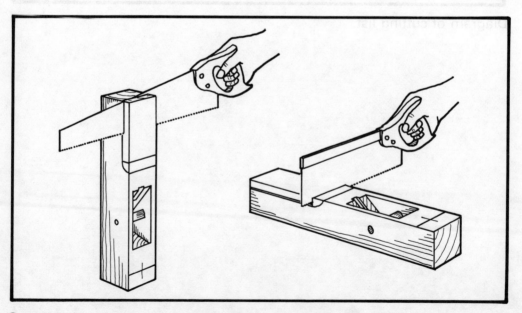

Saw out the waste square to the first line on the top of the stock.

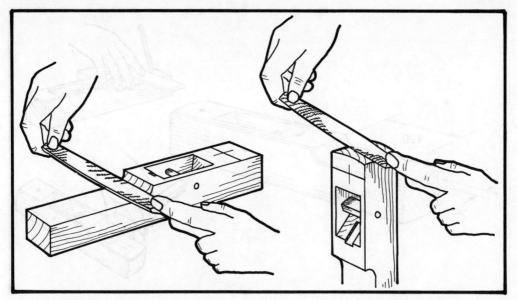

Shape the step of the stock with a chisel and a half round file or rasp, finish off with sandpaper.

Round off both ends either with a plane or a rasp and smooth the end grain with sandpaper.

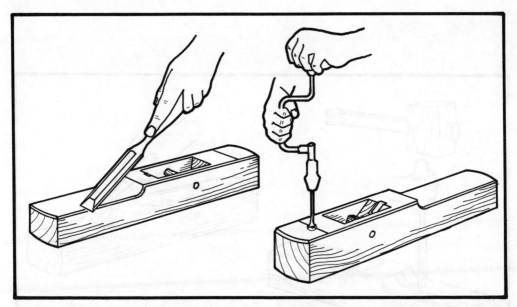

Use a chisel and a plane to cut a small chamfer all round the top edge of the stock.

With a carpenter's brace, drill a 12mm hole to a depth of 35mm in the front of the stock.

25

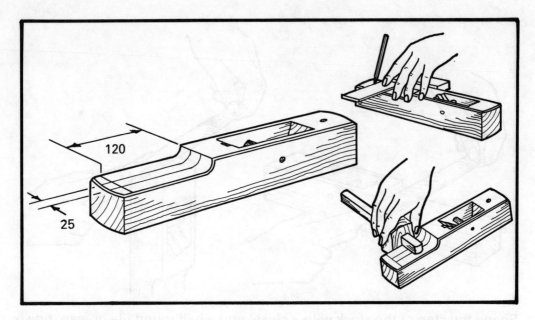

With a mortise gauge, mark the mortise for the handle in the centre of the stock. It should be 25mm wide and 120mm long.

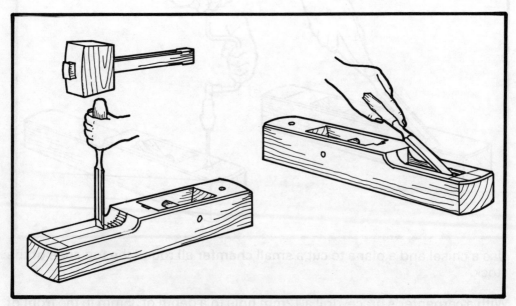

Chisel out the mortise to a depth of 6mm. Take care to keep the sides straight.

MAKING THE KNOB

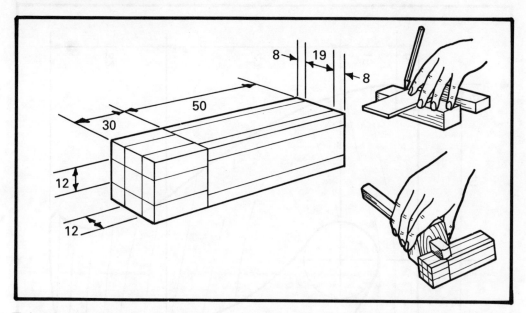

Take the timber prepared for the knob, and square a line all round 30mm from one end. Use a mortise gauge to mark a 12mm square tenon in the centre of the timber.

Gauge the other end of the knob with eight lines, 8mm in from each edge. This marks the eight-sided shape of the knob.

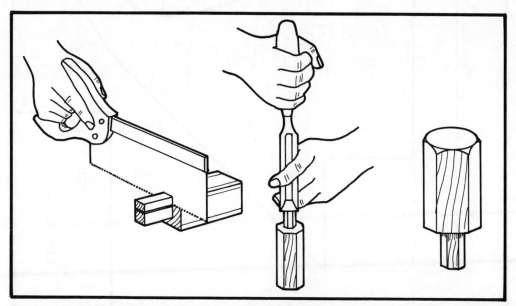

Use a tenon saw to cut the cheeks and shoulders of the tenon. Use a chisel to pare it down until it fits the 12mm hole in the stock.

Plane the corners of the knob down to the gauge lines. Smooth all the edges with sandpaper.

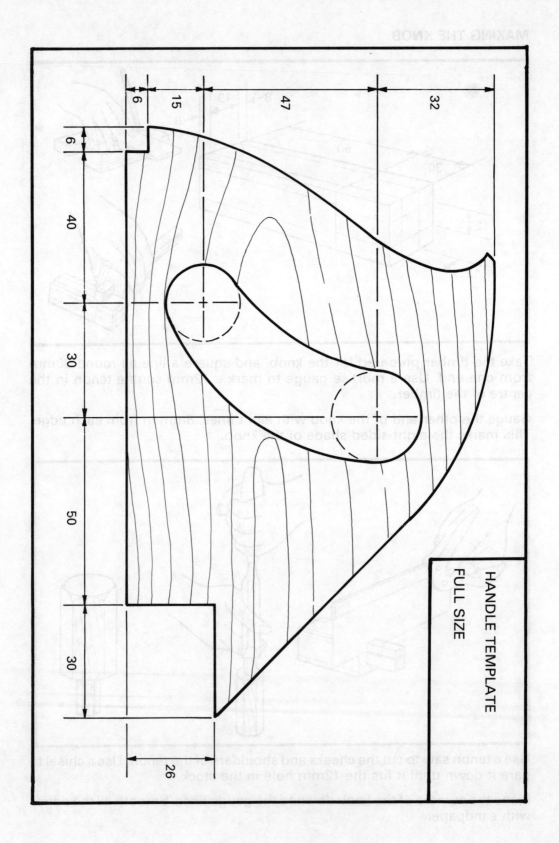

MAKING THE HANDLE

Trace the template on the opposite page and use this to mark the shape of the handle and the position of the holes.

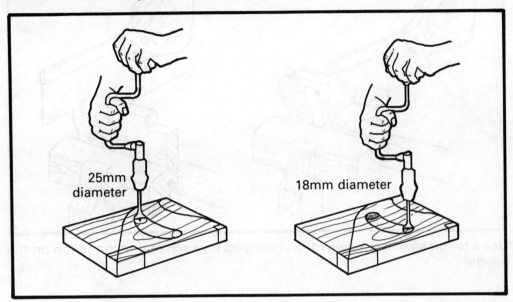

Use a 25mm bit to drill the hole at the top of the handle, and a 18mm bit for the hole at the bottom.

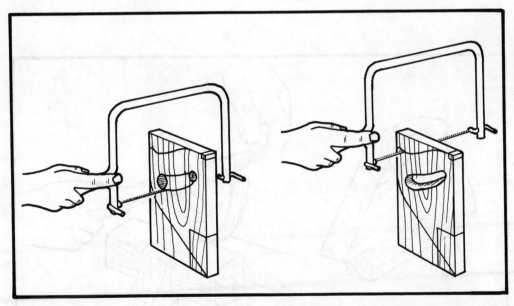

Use a bow saw or a coping saw to cut out the curved parts of the handle.

To saw out the hole, remove the saw blade from the frame, thread it through one of the holes in the handle, and fasten the blade to the frame again. The internal shape can now be cut.

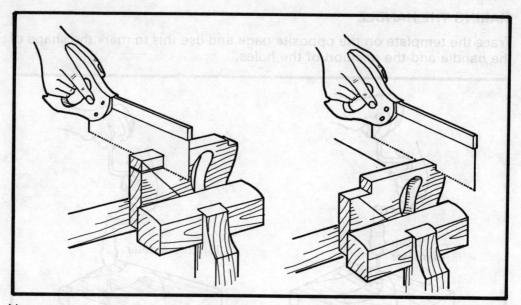

Use a tenon saw to cut away the shoulders on each end of the tenon on the handle.

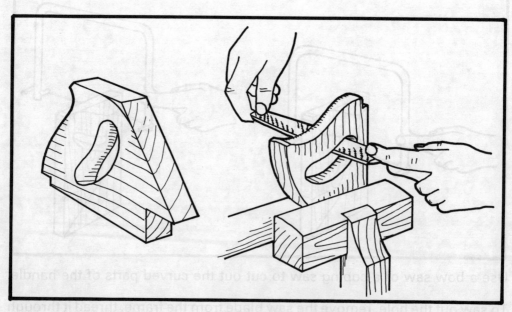

Cut the 45° slope on the front of the handle. Use chisels and rasps to round the 'grip' of the handle, and finish with sandpaper.

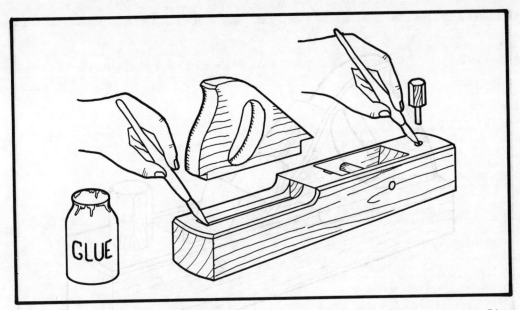

Make sure both the handle and the knob fit well into their mortises. Glue them in and clamp if necessary.

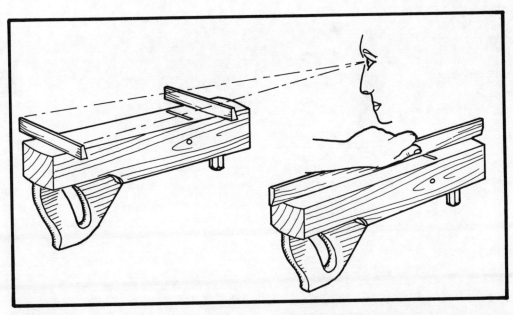

Plane the sole of the plane very carefully so that it is straight and free from winding.

The mouth opening should be no wider than 5mm.

Finished jack plane.

Rebate plane

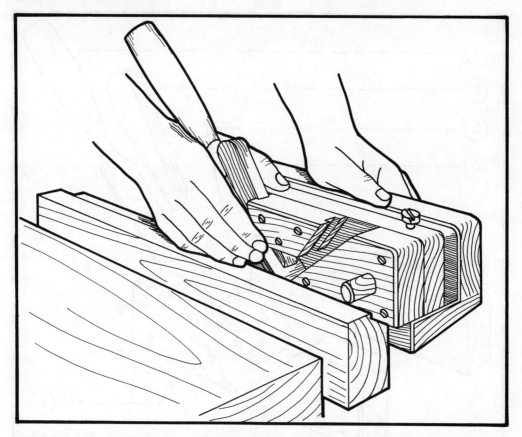

A rebate is a rectangular recess or step along the edge of a piece of wood. Cutting rebates is an essential part of all carpentry and joinery work; they are to be found in doors, door frames, windows and furniture. Rebates can be worked using saws or chisels, but this takes time and skill. A rebate plane is designed to do the job accurately and efficiently.

The tool described here uses a 20 or 25mm chisel as the cutter, which eliminates the problem of making or buying a blade specially for the plane, although one could be made from a piece of carbon steel.

The blade only extends to one side of the plane body, but because it has two blade beds the plane can be used in either direction. The chisel or cutter is held in place by a wedge and is adjusted in the same way as the wooden jack plane.

Rebate planes can be used without a fence, but for accurate and repetitive work an adjustable fence is very useful. The construction of two alternative versions are described on pages 46-56.

Do not begin to make this tool until you have obtained a suitable chisel or cutter, with a blade at least 100mm long.

LIST OF PARTS

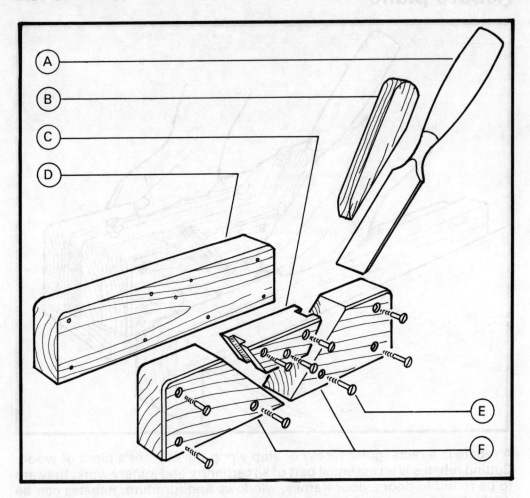

- **A** *Chisel or cutter.* This must be between 20 and 25mm wide and at least 100mm long.
- **B** *The wedge* fits into a groove in the bearing block, and holds the blade firmly to the bed of the plane.
- **C** *The bearing block* is screwed to the side piece of the plane, and holds the wedge in place.
- **D** *The side piece* is securely fastened to the three pieces of the stock.
- **E** *Screws.* Nine wood screws are used to fix the body of the plane together.
- **F** *The blade beds* as well as the bearing block are cut from one piece of timber. They must be exactly the same thickness as the width of the blade.

CUTTING AND PARTS LIST

Part	Name	Quantity, Material and Dimensions (mm)
A	Cutter	1 pc. Chisel, or tool steel 20-25 × 100
B	Wedge	1 pc. Timber 120 × 20 × 12
C & F	Stock	1 pc. Timber 250 × 65 × 'Dim A'
D	Side piece	1 pc. Timber 250 × 65 × 25
E	Screws	9 pcs. 1½" (37mm) No.8 Wood screws

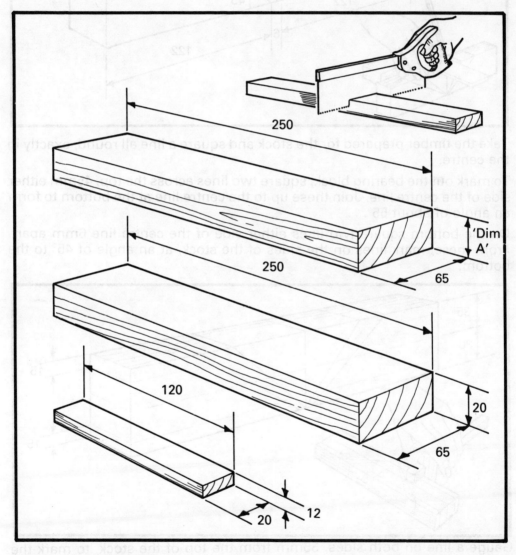

Diagram of cutting list.

'Dim A' refers to the width of the blade.

35

MAKING THE STOCK

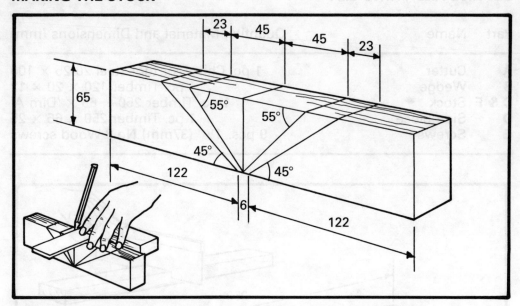

Take the timber prepared for the stock and square a line all round, exactly in the centre.

To mark out the bearing block, square two lines across the top, 45mm either side of the centre line. Join these up to the centre line at the bottom to form an angle of about 55°.

On the bottom square two lines either side of the centre line 6mm apart. From these, mark lines on the sides of the stock, at an angle of 45° to the bottom.

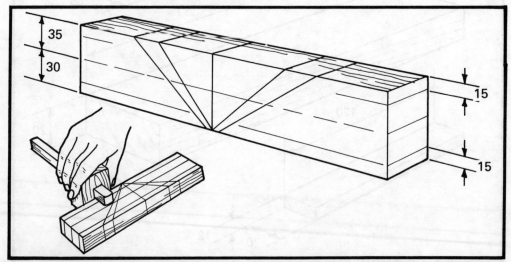

Gauge a line on both sides, 35mm from the top of the stock, to mark the bottom of the bearing block.

Gauge two lines all round 15mm from the top and bottom of the stock. These will mark the positions of the screws.

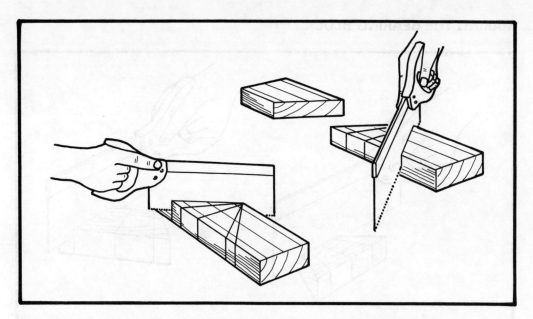

Saw the stock into the two blade beds with a tenon saw.

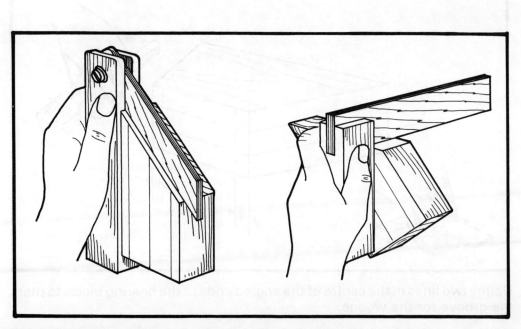

Plane the angled ends smooth, and check that the bed is square to the side and at an angle of 45° to the bottom of the stock.

MAKING THE BEARING BLOCK

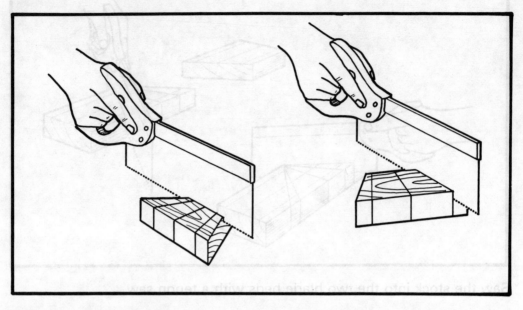

Take the centre part of the stock, cut it down to the gauge line, and saw the waste from both ends.

Gauge two lines in the centre of the angled ends of the bearing block, to mark the groove for the wedge.

Square lines across top and bottom 6mm from each edge, to mark the depth of the grooves.

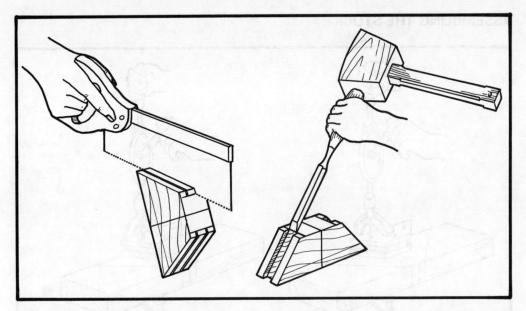

Use a tenon saw to cut the sides of the grooves, and remove the waste with a mallet and chisel.

MARKING OUT THE SIDE PIECE

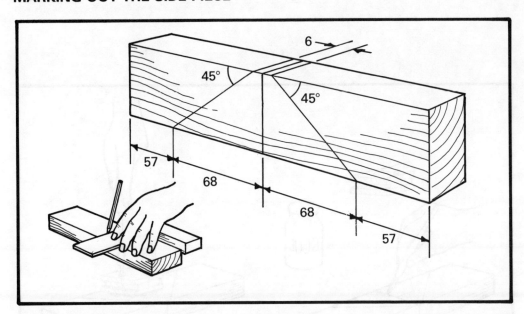

Take the timber prepared for the side piece and square a line all round exactly in the centre. On the bottom square two lines 3mm either side of the centre line.

On the top square two lines 68mm either side of the centre line. On the sides, join these points up to mark the position for the blade beds.

39

ASSEMBLING THE STOCK

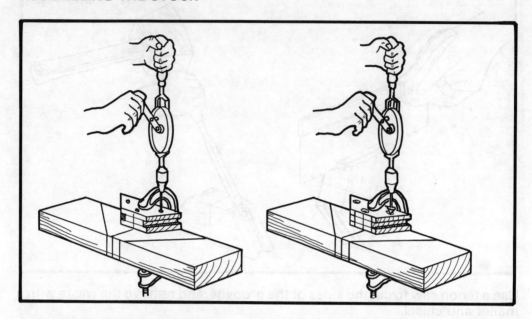

Cramp the wedge bearing block exactly in the centre of the side piece. Drill three clearance holes through the block, pilot holes into the side piece, and countersink for the screw heads.

Remove the cramps, apply glue and screw the wedge bearing block tightly in position.

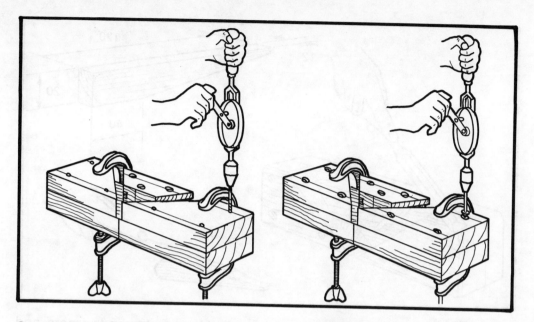

Cramp the two blade beds into position. Drill three clearance holes in each block, pilot holes into the side piece, and countersink for the screw heads.

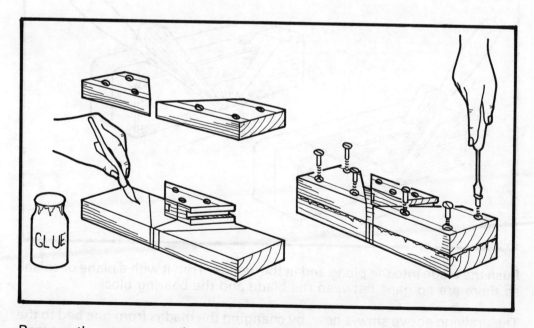

Remove the cramps, apply glue, and screw the two pieces tightly in position.

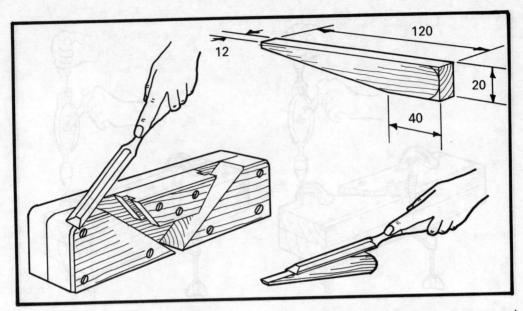

Plane the stock up square on all sides, round off the two top end corners, and use a plane or chisel to chamfer the top edges. This makes the tool more comfortable to hold. Leave the bottom edges square.

Cut and shape the wedge to the dimensions shown.

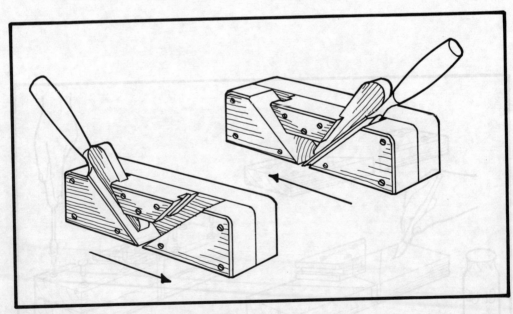

Push the blade into the plane and fit the wedge. Trim it with a plane or chisel, so there are no gaps between the blade and the bearing block.

The drawing above shows how, by changing the blades from one bed to the other, the plane can cut in both directions.

SETTING THE PLANE

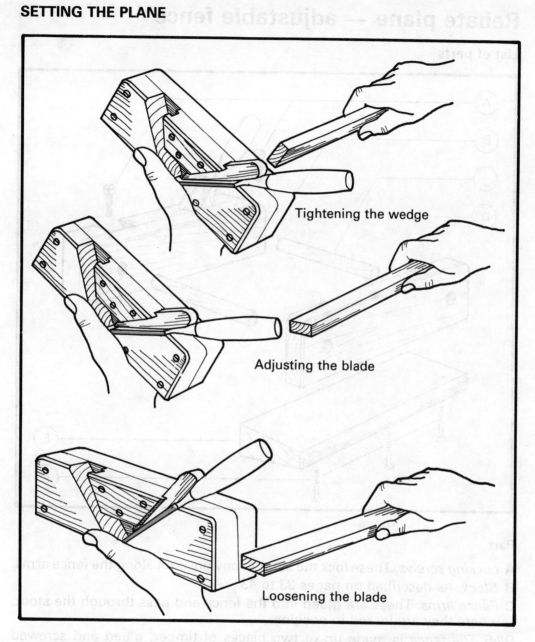

Tightening the wedge

Adjusting the blade

Loosening the blade

To tighten the wedge and the blade, strike the wedge with a piece of wood or a mallet.

To adjust the blade forward, strike the back of the cutter.

By hitting the back of the plane, you loosen the wedge and also bring the blade back.

43

Rebate plane — adjustable fence

List of parts

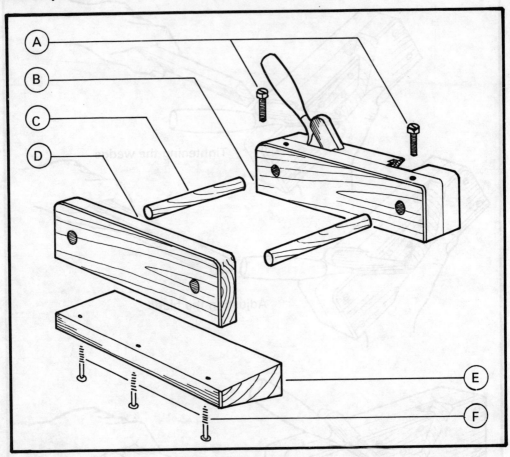

Part

A *Locking screws*. These lock the stock in any position along the fence arms.

B *Stock*. As described on pages 33 to 43.

C *Fence arms*. These are glued into the fence and pass through the stock, where they are locked in position.

D&E *The fence* is made up of two pieces of timber, glued and screwed together.

F *Wood screws*. Three are needed to fix the fence together.

CUTTING AND PARTS LIST

Part	Name	Quantity, Material and Dimensions (mm)
A	Locking screw	2 pcs. Bolts 25 × 6
C	Fence arms	1 pc. Timber 300 × 20 × 20
D & E	Fence	2 pcs. Timber 250 × 65 × 20
F	Screws	3 pcs. 1" (25mm) × 8 Woodscrews

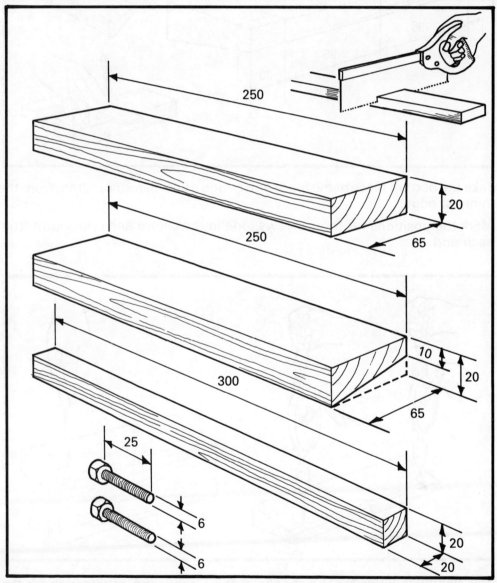

Diagram of cutting list.
One of the pieces for the fence can be gauged to a thickness of 10mm on one edge, and plane at an angle down to this line. This then becomes the bottom of the fence, and will make the tool lighter and easier to hold.

MAKING THE FENCE

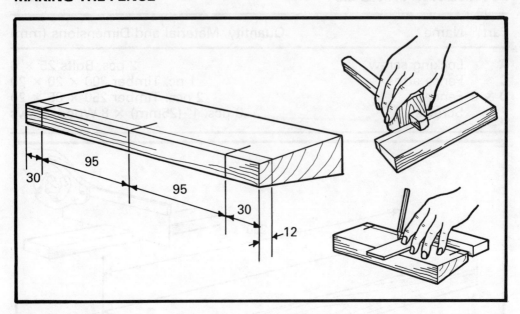

Take the bottom piece of the fence and gauge a line all round, 12mm from the thinnest edge.

Mark the positions for three screws, one in the centre and one 30mm from each end.

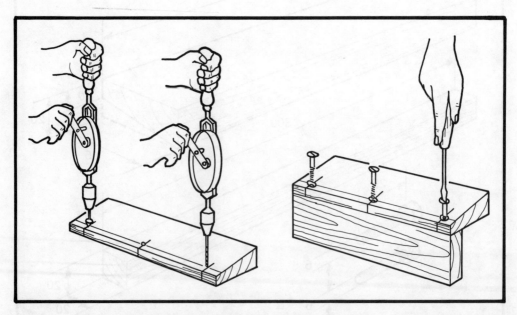

Drill clearance holes for the shanks of the screws and countersink for the heads. Clamp the two pieces together and drill pilot holes into the top of the fence.

Glue and screw the fence pieces tightly together.

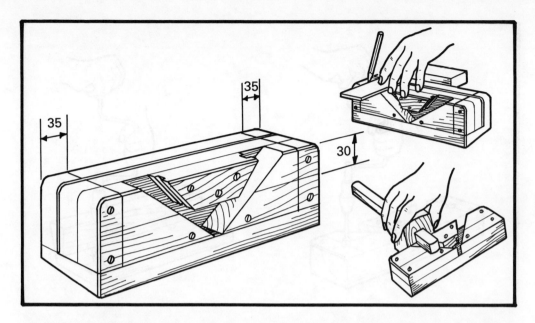

Put the plane stock and the fence together. Shape and plane the two parts to exactly the same size. To mark the position of the fence arms, keep the pieces together and square lines all round both pieces, 35mm in from each end. Gauge a line 30mm down from the top.

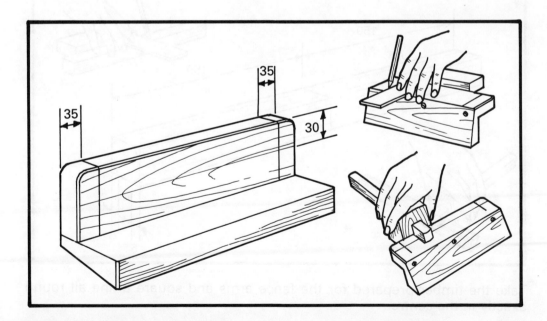

Separate the fence and the plane body and mark the inside faces of both pieces in the same way.

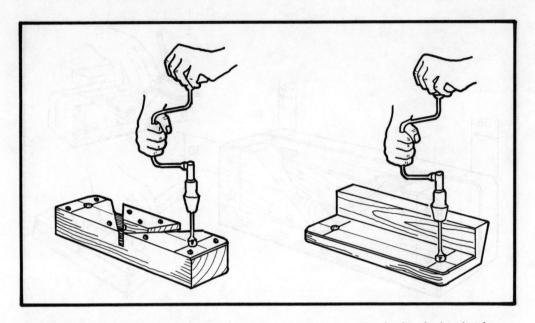

Use a 20mm bit and a carpenter's brace to bore out the holes in both pieces. Drill halfway through, from one side, turn the work over and finish the hole from the other side.

MAKING THE FENCE ARMS

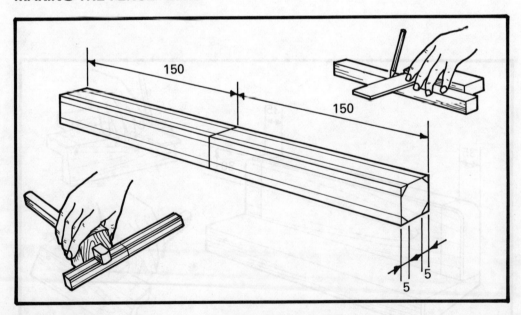

Take the timber prepared for the fence arms and square a line all round exactly in the centre.

Gauge eight lines 5mm in from each edge, and plane the corners off, down to the gauge lines, to form an eight sided section.

Saw the timber in half and fit both pieces into the fence, where they should be a tight fit.

With a plane or sandpaper carefully reduce the thickness of the parts that pass through the body of the plane, until they slide through easily.

Glue them into the fence, making sure they are square.

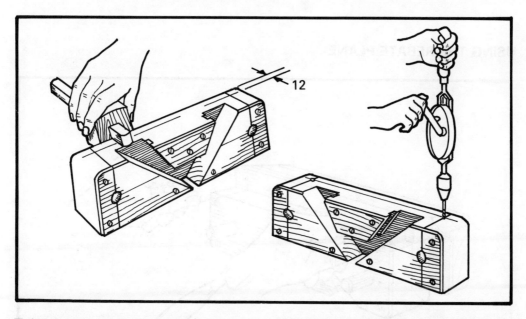

Take the plane, and gauge a line on the top of the side piece 12mm from the edge. Where this crosses the lines squared around for the fence arms, are the positions for the locking screws.

Use a 5mm bit and a wheel brace to bore two pilot holes for the locking screws, right through into the holes for the fence arms.

Take the two locking screws and saw slots in the heads with a hacksaw.

Carefully push the fence arms through the stock, screw the locking screws into the pilot holes (they will cut their own threads in the same way as a woodscrew). Leave the glue to dry for two or three hours.

USING THE REBATE PLANE

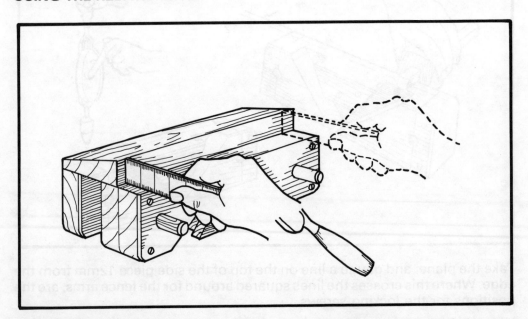

Set the fence with a ruler. It is important that the distance between the fence and the edge of the plane is the same at the front and the back.

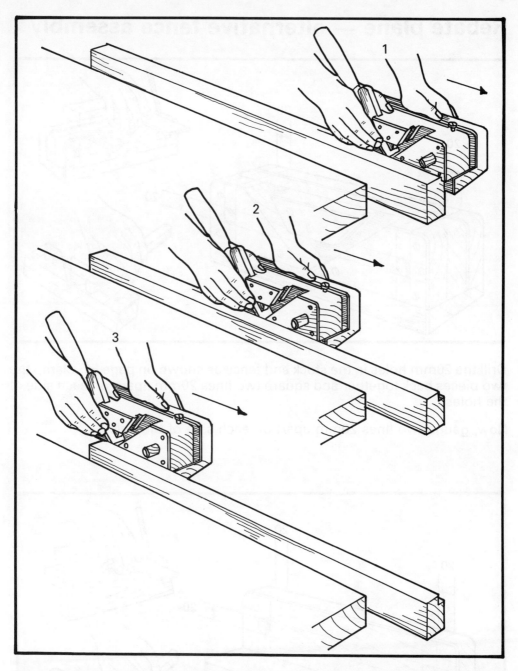

When cutting a rebate, make sure the blade is slightly proud of the side of the plane. Start at the far end and take one or two short shavings. As you proceed make longer and longer strokes, until you are cutting the whole length of the work piece. Finally trim the rebate to the desired depth, always maintaining a firm pressure on the fence. In this way the plane will cut a good square rebate. The numbered pictures show three stages in the process.

Rebate plane — alternative fence assembly

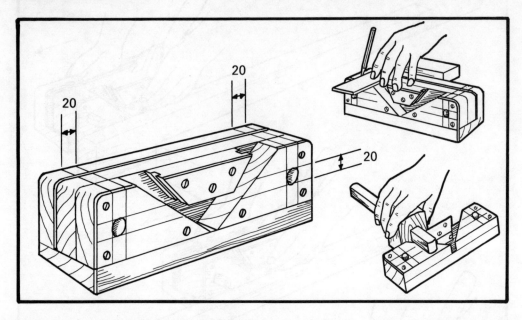

Drill the 20mm holes in the stock and fence as shown on page 50. Clamp the two pieces back together and square two lines 20mm apart, on each side of the holes.

Now, gauge two lines 20mm apart on each side of the holes.

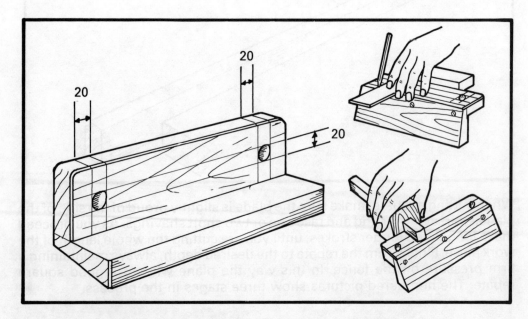

Mark the inside of the stock and the fence in the same way.

Square up the mortises with a mallet and chisel. Work halfway through from one side, turn the job over and finish off from the other side.

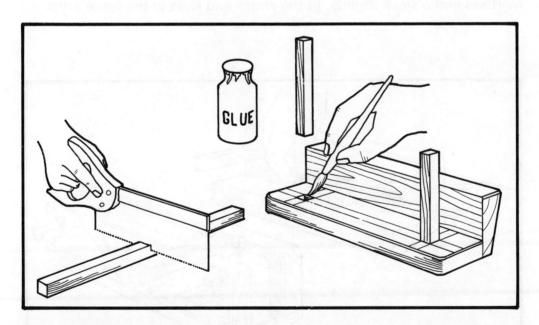

Take the timber prepared for the fence arms, leave it square in section, and saw it in half.

Fit and glue them into the fence. The mortises in the stock can be made bigger to allow the fence arms to slide through easily.

MAKING METAL SHOES

Metal shoes can be made from thin steel sheet. These will protect the fence arms from being bruised by the locking screws.

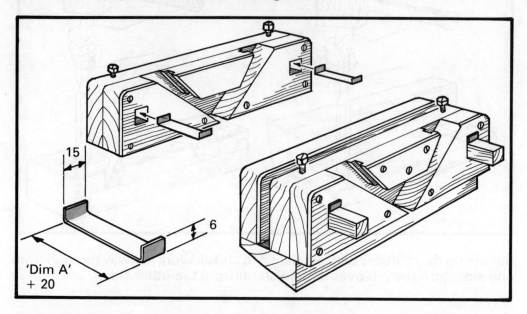

Cut and bend two pieces of steel to the dimensions shown. Enlarge the mortises in the stock slightly, fit the shoes and slide in the fence arms.

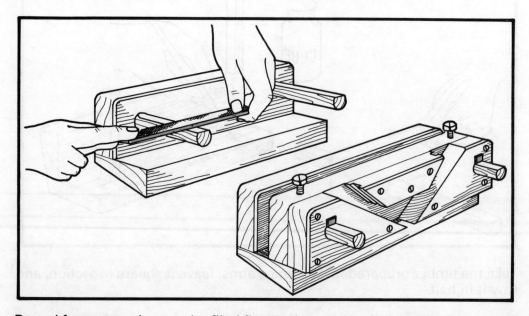

Round fence arms have to be filed flat on the top, to allow metal shoes to fit into the mortises.

Plough plane

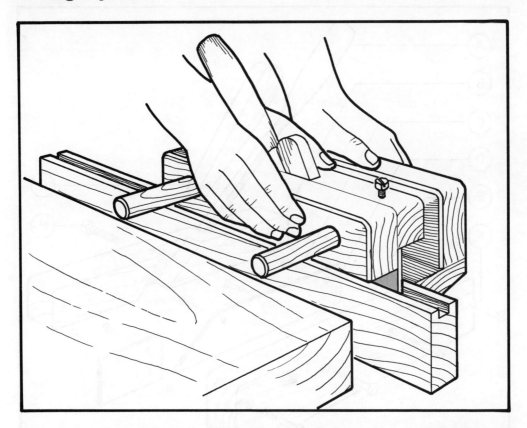

The plough plane is a difficult tool to make, because it involves both metalwork and woodwork. The metal sole plate has to be accurately made, and will be beyond the means of most rural carpenters, but with the help of a local metalworker this problem can be overcome.

The tool is capable of cutting grooves, for letting panels into frames, and rebates. Like the rebate plane described in the manual, it is designed to be used with a chisel as a cutter. The blade is held in position by a wedge and clamped tightly against the side piece of the body by a locking screw. This allows blades between 3 and 12mm in width to be used. Again they must be at least 100mm long. Do not begin to make the plane until you have found suitable blades.

The adjustable fence is a very necessary part of the tool and is made in exactly the same way as the fence for the rebate plane.

LIST OF PARTS

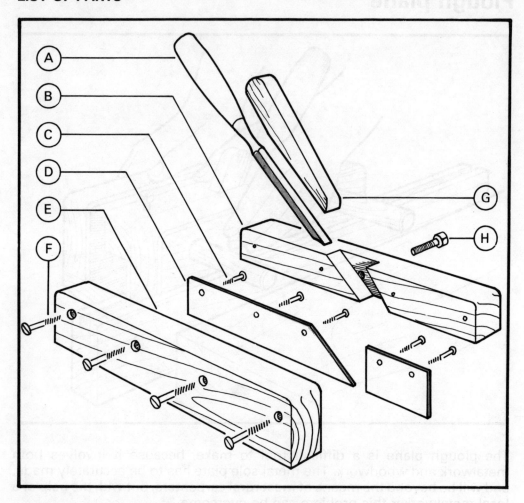

Part

A *Chisel* or *cutter*. This can be between 3mm and 12mm wide, and must be at least 100mm long.

B *The stock* is made from one piece of timber, with the blade bed and rebate for the sole plate cut out.

C *The sole plate* is made of one piece of mild steel plate 3mm thick.

D *Wood screws* are needed to secure the sole plate to the stock.

E *The side piece* is screwed to the stock and the blade is cramped tightly against it.

F *Woodscrews* are needed to secure the side piece to the stock.

G *The wedge* holds the blade in position.

H *The locking screw* allows blades of different widths to be used.

CUTTING AND PARTS LIST

Part	Name	Quantity, Material and Dimensions (mm)
A	Chisel or cutter	1 pc. Chisel or tool steel 100 long
B	Stock	1 pc. Timber 250 × 50 × 25
C	Sole plate	1 pc. Mild steel 250 × 35 × 3
D	Screws	5 pcs. Woodscrews 1″ (25mm) No.8
E	Side piece	1 pc. Timber 250 × 50 × 12
F	Screws	4 pcs. Wood screws 1¼″ (30mm) No.8
G	Wedge	1 pc. Timber 100 × 20 × 12
H	Locking screw	1 pc. Bolt 25 × 6

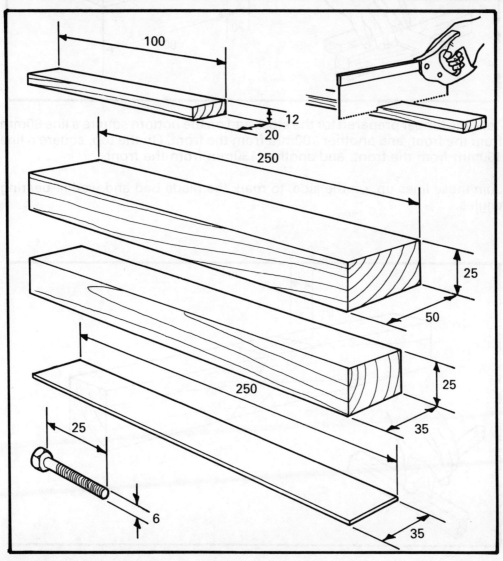

Diagram of cutting list.

MAKING THE STOCK

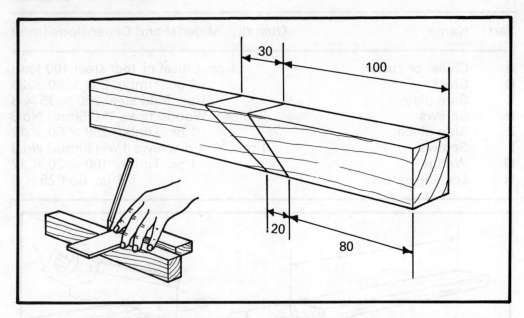

Take the timber prepared for the stock and on the bottom square a line 80mm from the front, and another 100mm from the front. On the top, square a line 100mm from the front, and another 130mm from the front.

Join these lines up, on the side, to mark the blade bed and wedge-bearing angles.

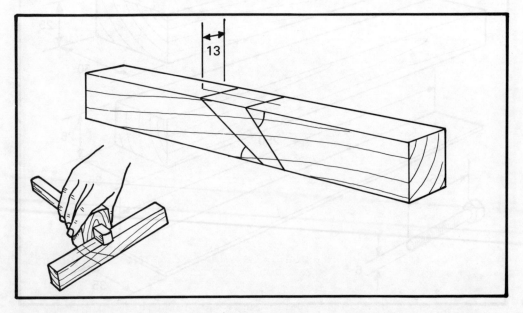

Gauge the depth of the recess for the blades to 13mm from the side.

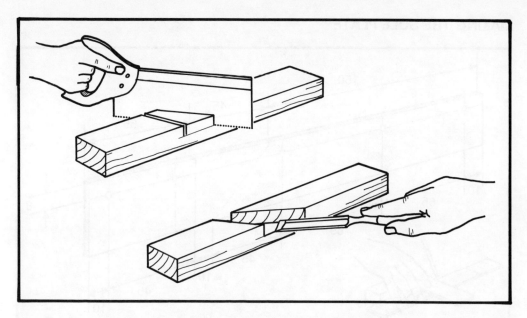

Use a tenon saw to cut the sides of the recess, and chisel out the waste, down to the gauge lines.

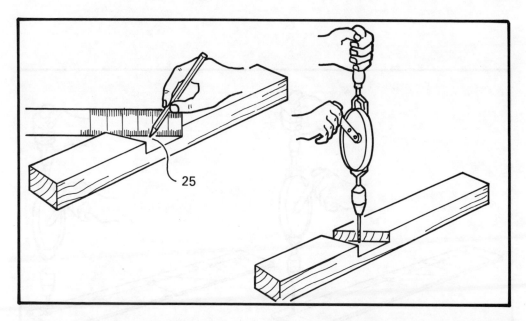

Use a pencil and ruler to mark a point, in the recess 25mm from the top of the stock, up against the blade bed.

From this point, drill a 5mm hole through the stock, right up against the blade bed. This is the pilot hole for the locking screw.

MAKING THE SOLE PLATE

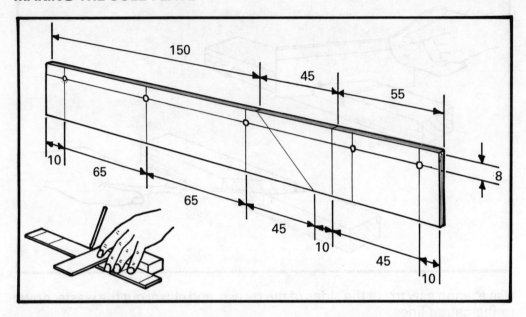

Take the mild steel plate and mark it out as shown. There is a full size template at the back of the manual.

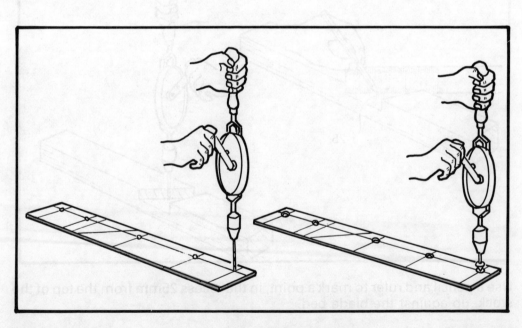

Drill five clearance holes for No.8 screws, at the top of the plate, and countersink to let the screw heads right in.

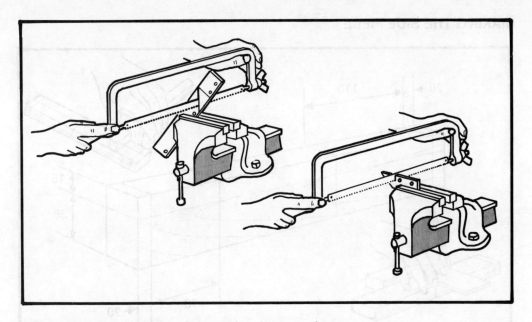

Hacksaw the two parts of the sole plate to shape.

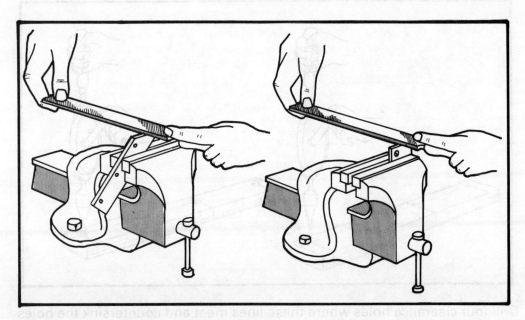

File all the edges square and straight.

MAKING THE SIDE PIECE

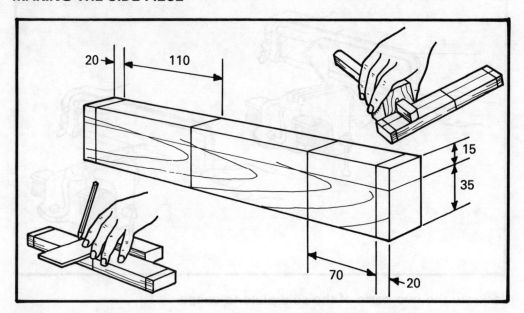

Take the timber prepared for the side piece and square a line 20mm from each end. Square a line 70mm from one end, and another 110mm from the other end.

Gauge a line 20mm from the top of the stock.

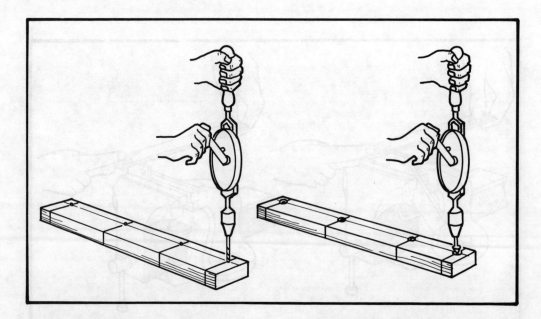

Drill four clearance holes where these lines meet and countersink the holes for the screw heads.

ASSEMBLING THE STOCK

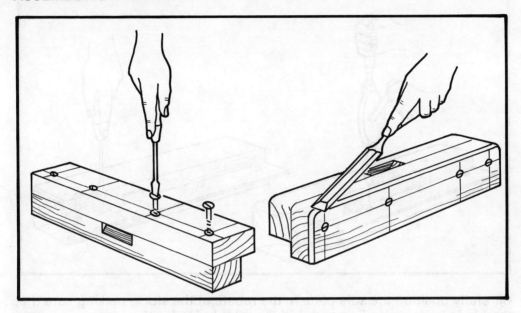

Clamp the side piece to the stock and drill pilot holes for the screw threads. Screw the stock and the side piece tightly together.

Smooth the two parts of the stock with a plane, and finish by rounding off the two top corners and chamfering the edges so the plane is comfortable to hold. Leave the bottom edges square.

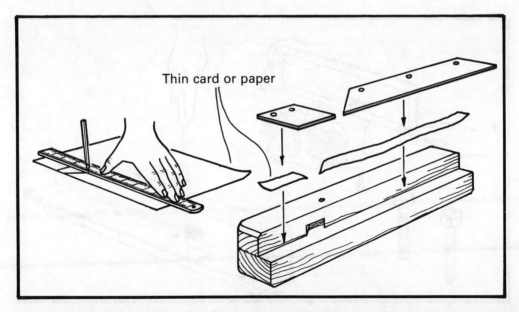

To give the sole plate clearance in the groove cut by the plane, a thin piece of card or paper should be used as packing between the stock of the plane and the sole plate.

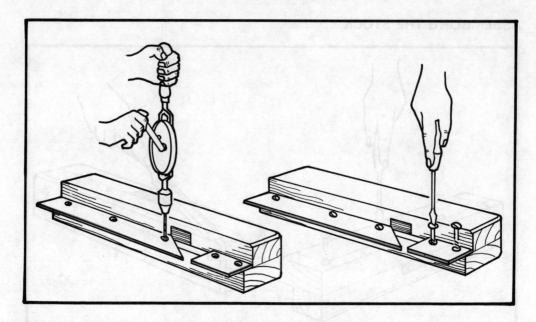

Carefully position the sole plate in the rebate of the stock, making sure that the blade beds are in line and drill the pilot holes for the screws.

Screw the sole plate securely onto the stock.

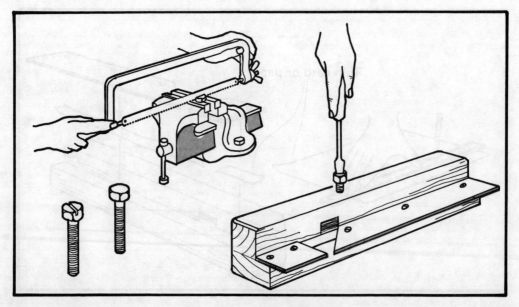

Use a hacksaw to cut a slot in the top of the locking screw, and screw it into the pilot hole in the side of the stock. It will cut its own thread in the same way as a wood screw.

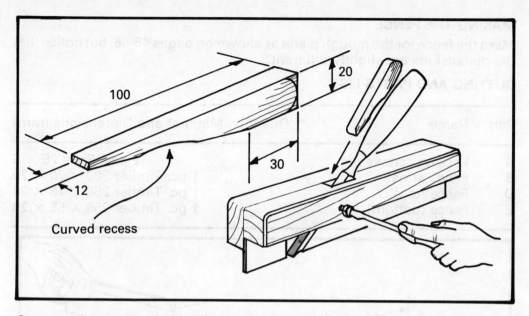

Cut and shape the wedge as shown; the curved recess at the bottom is to allow for the locking screw.

Place the chisel or blade into the stock, making sure the wedge fits well between the blade and the stock. Clamp the blade against the side piece with the locking screw.

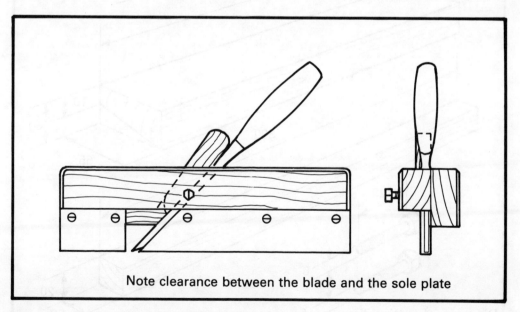

Drawing of the finished plough plane stock.

MAKING THE FENCE

Make the fence for the plough plane as shown on pages 46-56, but notice that the dimensions are slightly different.

CUTTING AND PARTS LIST

Part	Name	Quantity, Material and Dimensions (mm)
A	Locking screws	2 pcs. Bolts 25 × 6
B	Fence arms	1 pc. Timber 300 × 20 × 20
D	Fence (top)	1 pc. Timber 250 × 75 × 20
E	Fence (bottom)	1 pc. Timber 250 × 55 × 20

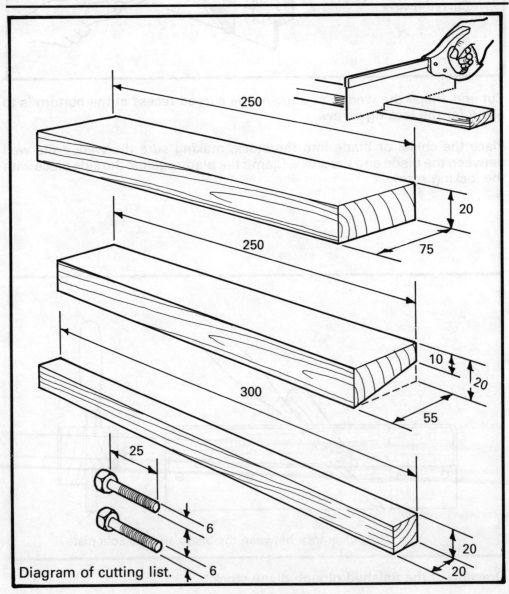

Diagram of cutting list.

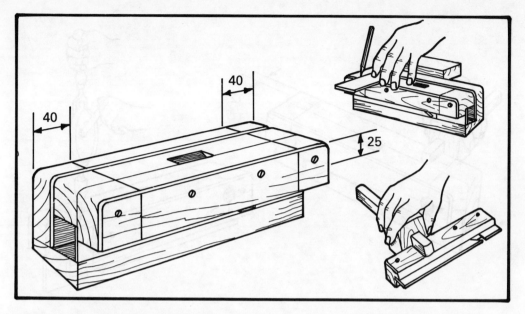

To mark the position of the fence arms, clamp the stock and fence together, square lines across the top and sides 40mm from each end. Gauge two lines 25mm from the top.

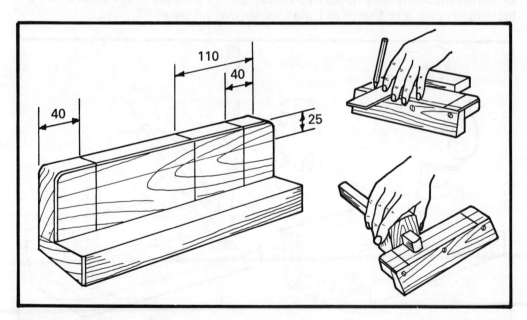

Separate the fence and the plane body, and mark the inside faces of both pieces in the same way.

Bore the 20mm holes, and fit the fence arms as shown on pages 50 and 51.

Also, on the top fence, square a line all round 110mm from the front end.

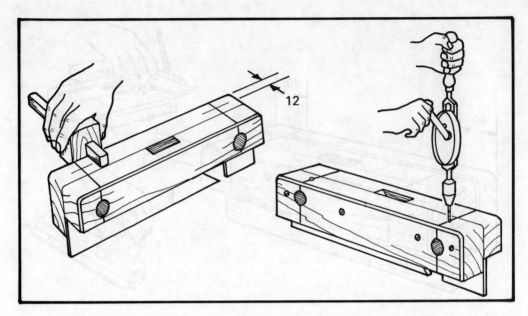

Gauge a line on the top of the stock 12mm from the side. Where this line crosses the squared pencil lines, mark the positions for the locking screw pilot holes.

Drill the two 5mm pilot holes right through to the fence arm holes. Drive the locking screws into the pilot holes as shown on page 52.

Drill a 20mm hole through the top of the fence. This will allow the blade locking screw to be tightened when the fence is in position.

The plough plane is used in exactly the same way as the rebate plane: follow the directions on pages 50 and 51.

68

Spokeshave

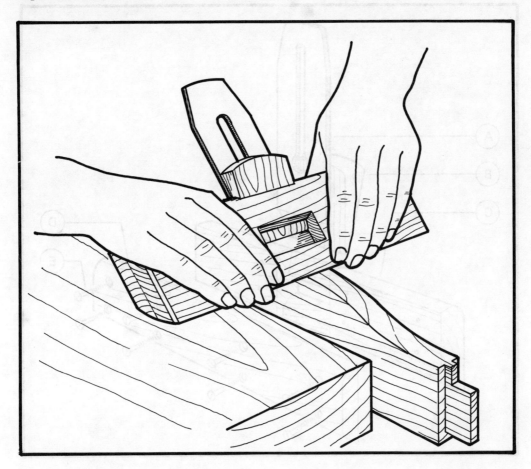

A spokeshave is used for smoothing curved surfaces on decorative work, where a smoothing plane cannot be used. It is not an essential tool, but used with a bow saw or a coping saw, it will improve a carpenter's range of products and his quality greatly. It is ideal for shaping curved parts of chairs, table legs and head boards.

The stock is made from two pieces of timber screwed together, with a recess cut out of the inside of the front piece for the wedge and blade. The blade is wedged into the centre of the stock, and the shavings pass through a window in the front of the stock.

It is the easiest of the planes to make, and as this design uses an ordinary jack plane blade, very little time or money needs to be spent on it.

LIST OF PARTS

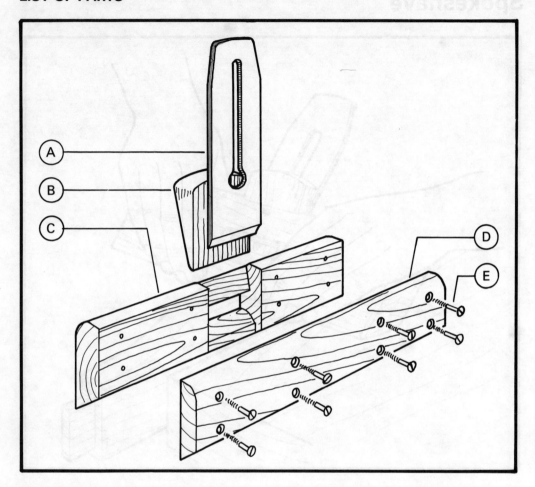

Part

A *The blade.* Either a plane blade or a spokeshave blade can be used in this tool.

B *The wedge* holds the blade securely in the stock.

C *The front stock* has a recess cut out to take the blade and the wedge, and a window for the shavings to pass through.

D *The back stock.* This is screwed to the front stock, and the blade is wedged up to it.

E *Screws* are needed to fix the two parts of the stock together.

CUTTING AND PARTS LIST

Part	Name	Quantity, Material and Dimensions (mm)
A	Blade	1 pc. Plane or spokeshave blade
B	Wedge	1 pc. Timber 100 × 'Dim A' × 15
C	Front stock	1 pc. Timber 250 × 85 × 20
D	Back stock	1 pc. Timber 250 × 65 × 15
E	Screws	8 pcs. Wood screws 1" (25mm) No.8

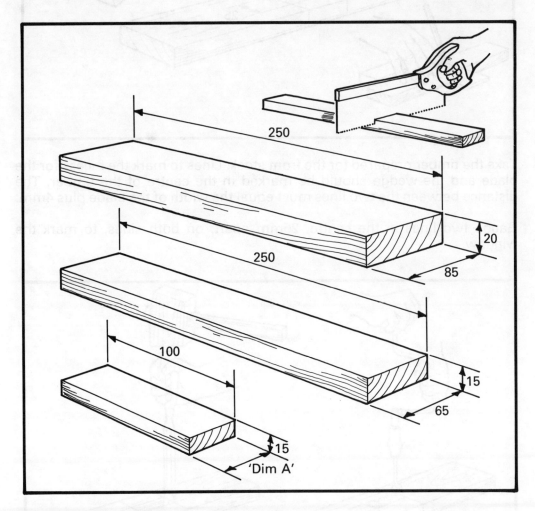

'Dim A' refers to the width of the blade.

Diagram of cutting list.

71

MAKING THE STOCK

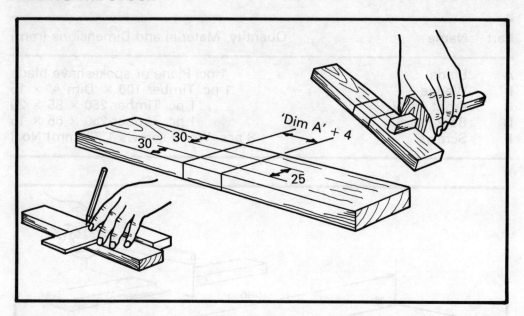

Take the timber prepared for the front stock. Lines to mark the recess for the blade and the wedge should be marked in the centre of the timber. The distance between the two lines must equal the width of the blade plus 4mm.

Gauge two lines in the centre, 25mm apart, on both sides, to mark the window.

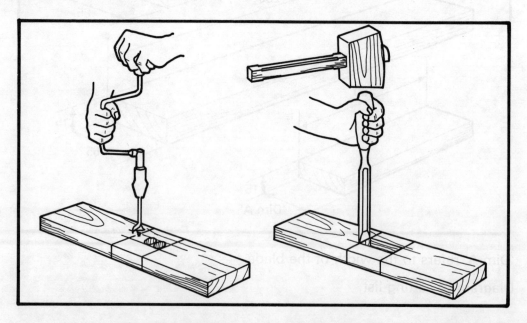

Use a brace and bit to drill a number of holes, to ease the cutting of the mortise or window. Finish off the mortise with a mallet and chisel.

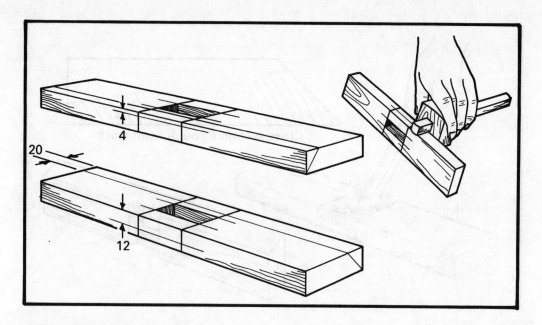

Gauge a line on one side 20mm from one edge. Mark a 45° angle on both ends for the sole.

On the edge marked as the sole, gauge the depth of the recess to 4mm. On the opposite edge, mark the depth to 12mm.

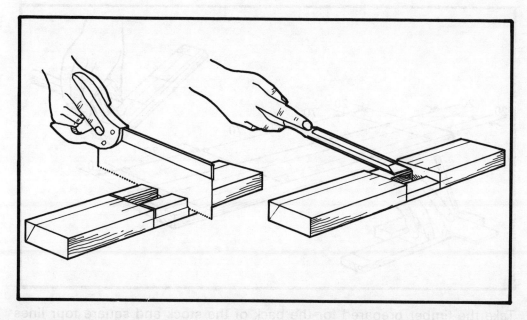

Saw the sides of the recess carefully down to the gauge lines. Remove the waste with a chisel. This angled surface then becomes the bearing for the wedge.

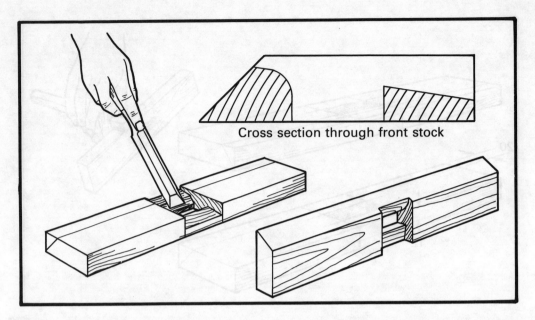

Cross section through front stock

Use a chisel to round off the inside edge of the window nearest the sole. This allows the shavings to pass easily through the stock.

Plane the sole to an angle of 45° to the front of the stock.

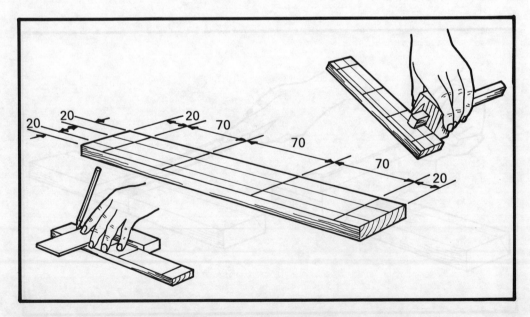

Take the timber prepared for the back of the stock and square four lines across, two 20mm from each end and two 70mm from each end. Gauge two lines 20mm from each edge. Where these lines meet marks the holes for eight screws.

74

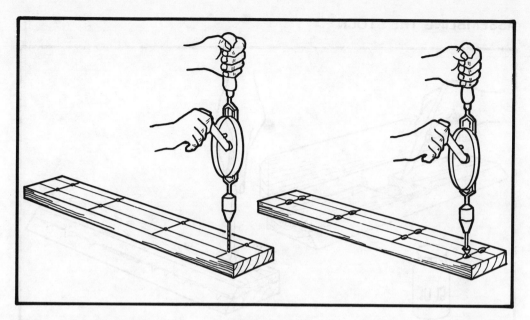

With a wheel brace and bit, drill eight clearance holes, and countersink them for the screw heads.

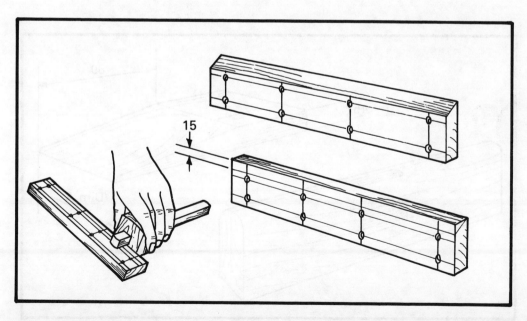

Gauge a line 15mm from one edge, on the same side as the countersunk holes, and mark a 45° angle on both ends. Plane the angled sole down to the gauge line.

ASSEMBLING THE STOCK

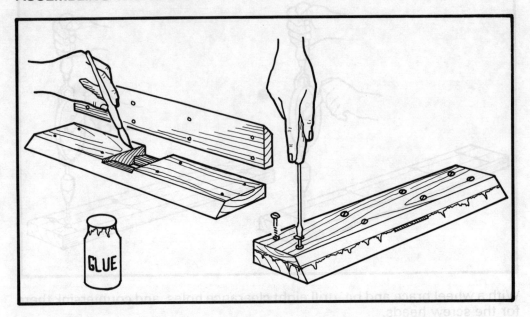

Clamp the back and the front stock together, and drill pilot holes for the screw threads.

Remove the clamp and glue and screw the two pieces together.

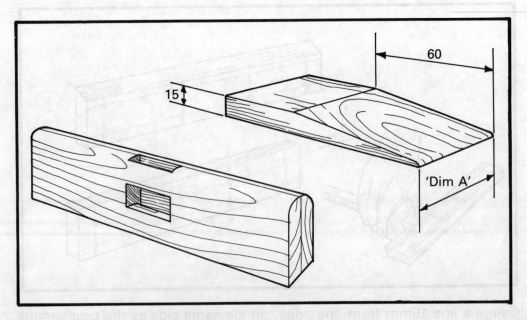

Round off the top of the stock so it is comfortable to hold. Cut and shape the wedge to the dimensions shown.

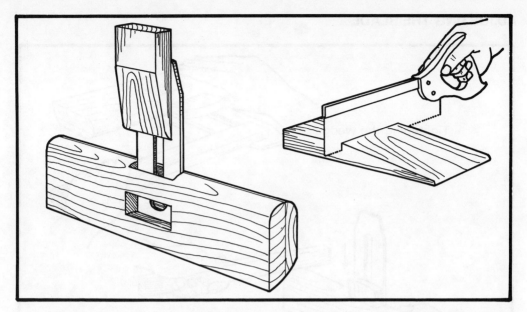

Push the blade into the stock, and fit the wedge between the blade and the bridge of the stock, making sure there are no gaps between the two parts.

Cut the end of the wedge off, about 60mm from the sharp end.

USING THE SPOKESHAVE

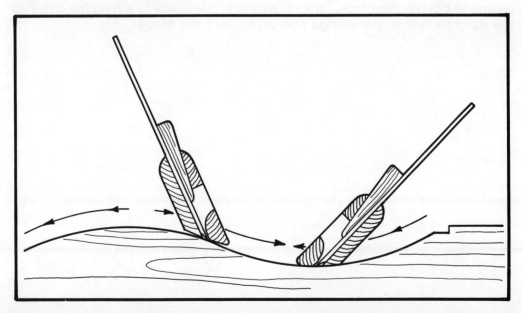

When using the spokeshave, it is important to work with the grain. This means that the spokeshave will often have to be used in both directions when smoothing one curve.

ADJUSTING THE BLADE

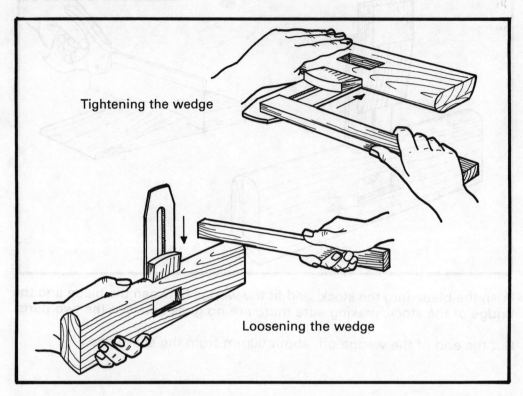

Tightening the wedge

Loosening the wedge

This spokeshave is adjusted in exactly the same way as the planes. Strike the wedge to tighten the blade, strike the back of the blade to adjust the cut. Tap the stock to loosen the wedge and bring back the blade.

Sash cramp

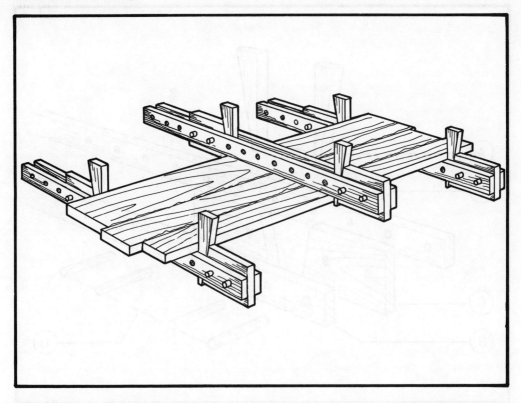

Sash cramps are essential tools for all joinery and furniture making. Every workshop should have at least two, but when it comes to gluing up large table tops, or frames with more than two rails, the more sash cramps you have available the easier the work becomes.

The whole cramp is made of timber. The workpieces are tightened together with a pair of wedges working against angled jaws. The jaws are fixed at any point along the bars using four wooden dowels.

These cramps are capable of giving adequate pressure for gluing up timbers 25mm thick. Extra care should be taken when cutting joints in timber thicker than this, so the pieces go together with little pressure. Always remember to slip a piece of paper or a wood shaving between the bars and the joint to prevent them being glued together by mistake.

LIST OF PARTS

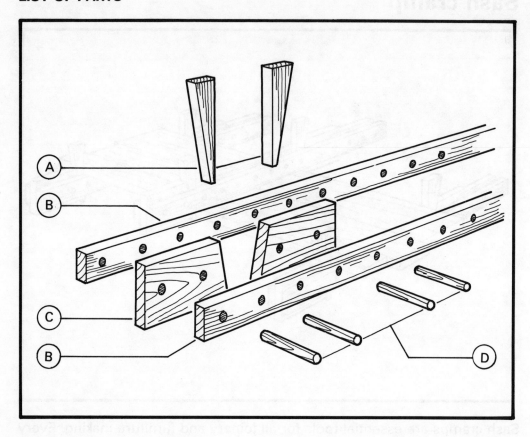

Part

A *Wedges.* As the wedges are driven across the angled jaws, they put pressure on the workpiece.

B *The bars.* Can be any length up to 2.5 metres, and are drilled with holes every 100mm.

C *The jaws.* Each have two holes bored through 100mm apart. They are sandwiched by the bars.

D *Dowels.* These allow the jaws to be fixed at any point along the length of the bars.

CUTTING AND PARTS LIST

Part	Name	Quantity, Material and Dimensions (mm)
A	Wedges	1 pc. Timber 250 × 45 × 20
B	Bars	2 pcs. Timber * × 75 × 20
C	Jaws	1 pc. Timber 425 × 175 × 20
D	Dowels	1 pc. Timber 400 × 20 × 25

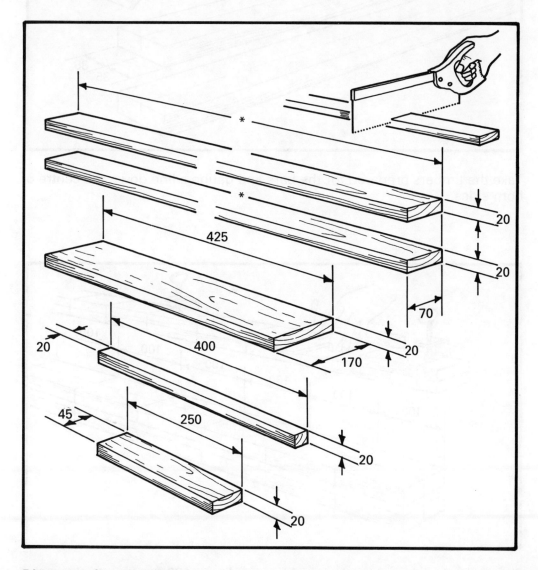

Diagram of cutting list.

*It is up to the reader to decide on the length of the bars.

MAKING THE BARS

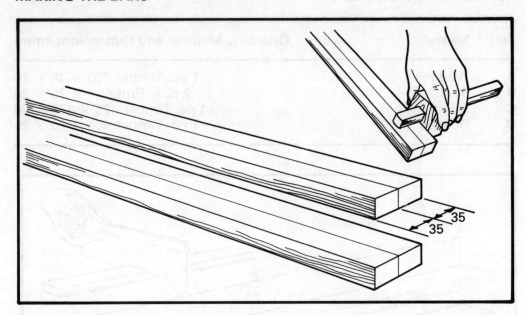

Take the timbers prepared for the bars and gauge a line down the centre of both sides.

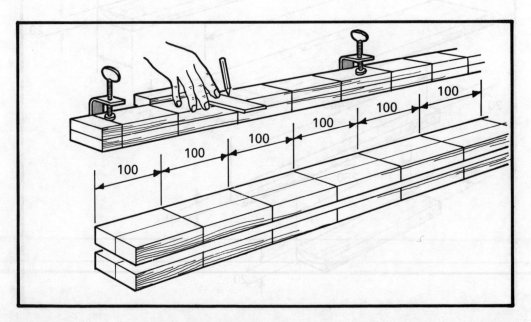

Clamp the two bars together and square lines, 100mm apart, along the whole length.

Remove the clamps and mark these lines on all four sides.

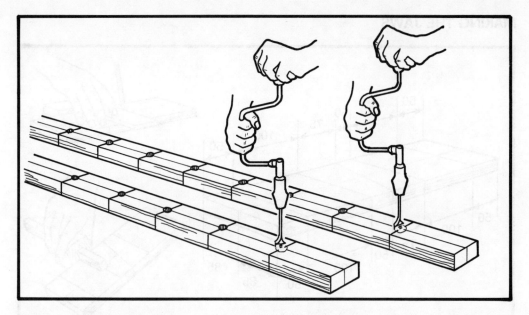

With a carpenter's brace and a 25mm bit, drill holes where the lines meet. Remember to drill halfway through from one side, turn the work over and finish the hole off from the other side.

MAKING THE DOWELS

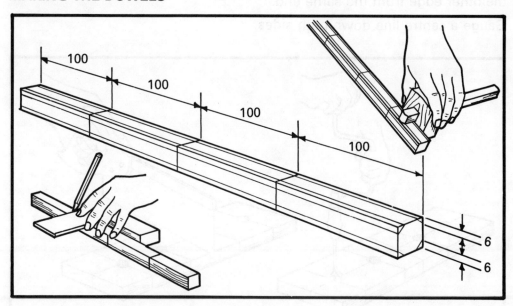

Take the timber prepared for the dowels, and square lines all round every 100mm.

Gauge eight lines 6mm in from each edge. Plane the timber to an eight-sided section. Saw it into four pieces, and plane them to fit into the holes in the bars.

MAKING THE JAWS

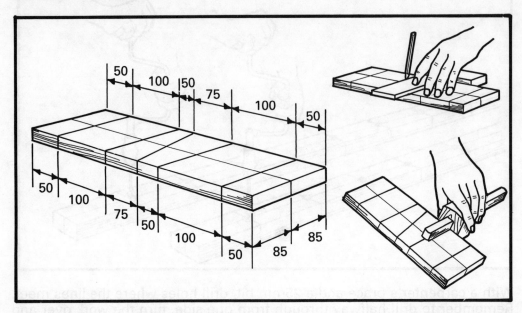

Take the timber prepared for the jaws, and square lines all round 50mm from both ends, and 150mm from both ends. Mark the angled ends of the jaws by measuring and marking, on one edge, 200mm from one end, and 225mm on the other edge from the same end.

Gauge a centre line down both sides.

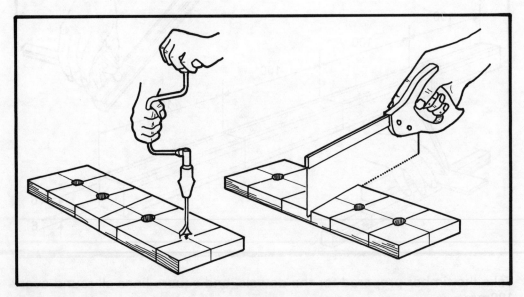

Drill four 25mm holes where the squared lines meet the gauged line. Drill halfway through from both sides.

Use a tenon saw to cut the angled ends of the jaws. Plane the end grain up square to the sides.

MAKING THE WEDGES

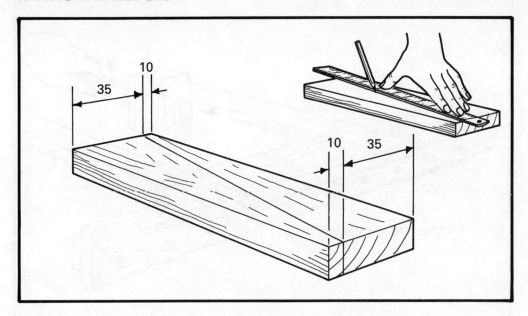

Take the timber prepared for the wedges. Square lines on the end grain of opposite corners, 10mm from each edge. Join these up on the sides with a ruler and pencil.

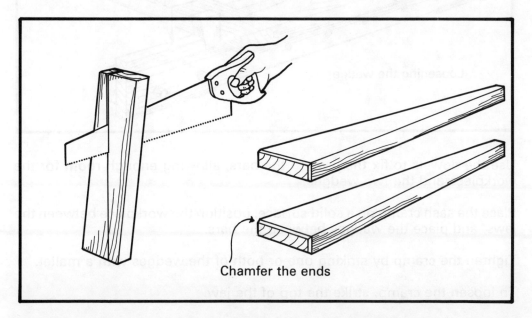

Chamfer the ends

Saw the wedges to shape and chamfer the end grain. This will prevent the wedges splitting when they are struck with a mallet.

USING THE SASH CRAMPS

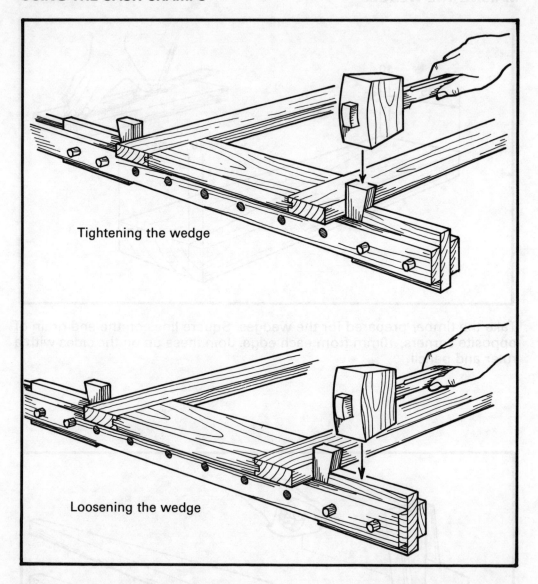

Tightening the wedge

Loosening the wedge

Use the dowels to fix the jaws to the bars, allowing enough room for the workpiece and the two wedges.

Place the sash cramp on a solid surface, position the workpiece between the jaws, and place the wedges between the bars.

Tighten the cramp by striking one or both of the wedges with a mallet.

To loosen the cramp, strike the top of the jaw.

When gluing up, remember to place a piece of paper between the workpiece and the bars, to prevent them being glued together.

Bench cramp

Small bench cramps can be made along the same lines as the sash cramp described earlier. As a substitute for metal gee cramps, they can be used for gluing up small pieces of joinery, or clamping timber to the workpiece.

In this design the bars are jointed into the jaws. One jaw is fixed with woodscrews, while the other is free to move between the bars, but can be locked in a number of positions with a dowel. Different sized cramps may be more suitable for the work you intend to do, so use the dimensions given here only as a guide.

Two or three of these cramps would be adequate for a small workshop, but more can easily be made if the need arises.

LIST OF PARTS

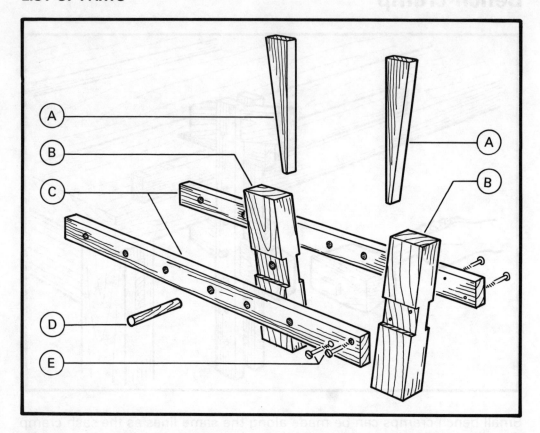

Part

A *Wedges.* As the wedges are driven across the angled jaws, they put pressure on the workpiece.

B *The jaws* have sockets cut out of each side to let the bars in.

C *The bars* are drilled with holes 80mm apart, and are screwed to one of the jaws.

D *The dowel* allows the moving jaw to be fixed in a number of positions along the bar.

E *Screws.* Four wood screws are needed to secure the bars to the fixed jaws.

CUTTING AND PARTS LIST

Part	Name	Quantity, Material and Dimensions (mm)
A	Wedges	1 pc. Timber 250 × 45 × 20
B	Jaws	1 pc. Timber 510 × 70 × 45
C	Bars	2 pcs. Timber 500 × 50 × 20
D	Dowel	1 pc. Timber 100 × 20 × 20
E	Screws	4 pcs. Wood screws 2" (50mm) No.8

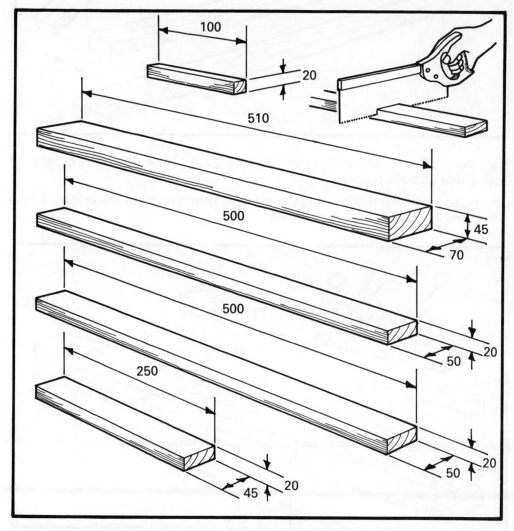

Diagram of cutting list.

MAKING THE BARS

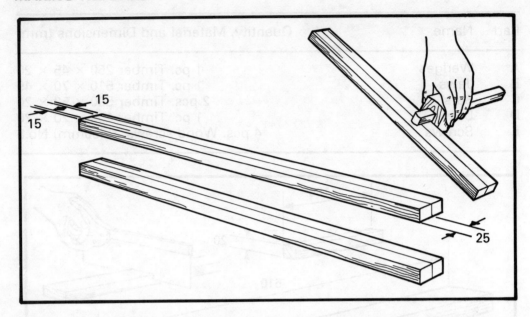

Take the timbers prepared for the bars and gauge lines down the centre, on both sides of both pieces.

On one end of each piece, gauge two lines, 15mm in from the edges. These will mark the position of the screws.

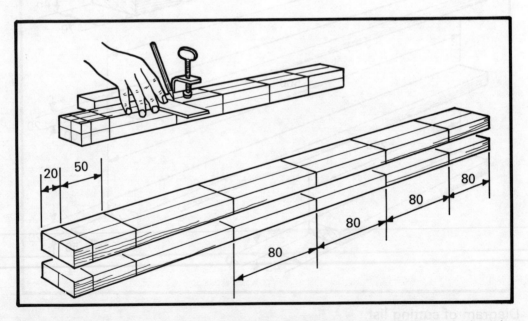

Clamp the bars together, and square two lines across for the screw holes, the first 20mm from the end, the second 70mm from the end.

To mark the dowel holes, start at the other end, and measure 80mm in for the first hole. Square three more lines all round each 80mm apart.

Use a carpenter's brace and a 20mm bit to bore the dowel holes. Remember to drill halfway through from one side, turn the work over and finish the hole from the other side.

Drill clearance holes for two screws on each bar on opposite sides. Arrange the holes so that the screws will not meet in the centre of the jaw, and countersink them for the screw heads.

MAKING THE JAWS

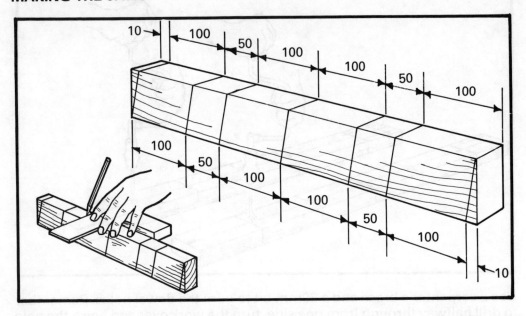

Take the timber prepared for the jaws and square lines across one edge. The first line, 10mm from the end, the second 110mm, the third 160mm and the fourth 260mm, the fifth 360mm from the end, the sixth 410mm and this should be 100mm from the other end.

Do exactly the same on the other edge but begin marking from the other end. Join these lines up on the sides to mark the angled ends of the jaws.

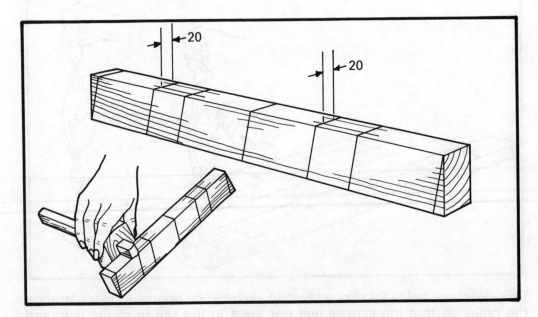

Gauge the depth of the four sockets to 12mm using a marking gauge.

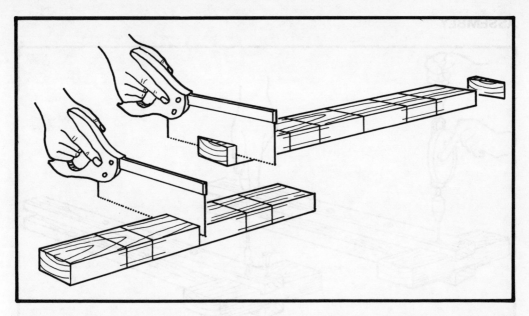

Use a tenon saw to cut the angles on the end of the timber and separate the two jaws by sawing down the centre line.

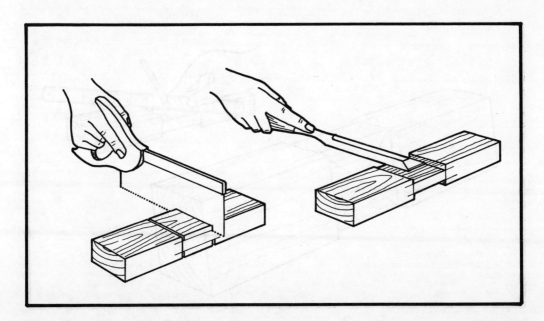

Saw the sides of the sockets on both jaws, and remove the waste, down to the gauge line, with a chisel.

ASSEMBLY

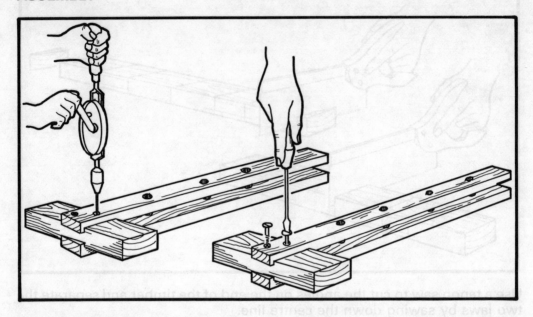

Push the two bars into the socket in one of the jaws, and drill pilot holes for the screw threads.

Glue and screw the parts tightly together.

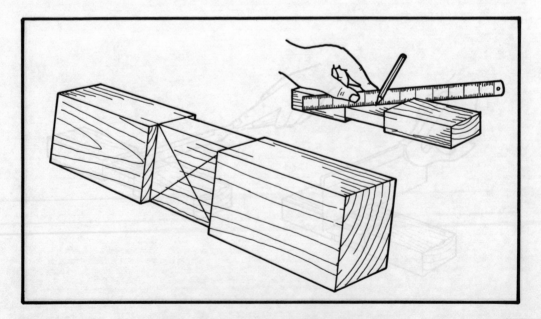

Take the remaining jaw and mark the diagonals, on the bottom of both sockets.

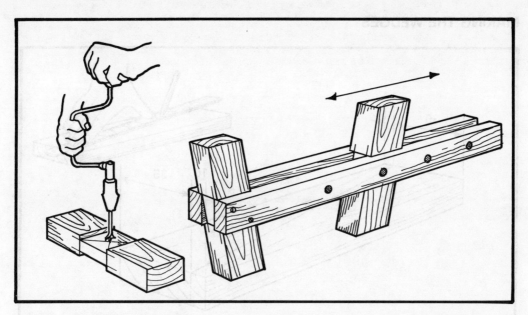

Drill a 20mm hole exactly where the diagonal lines meet. Fit and slide the moving jaw between the two bars.

MAKING THE DOWEL

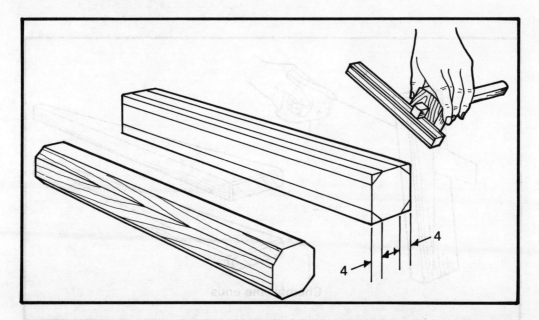

Take the timber prepared for the dowel and gauge eight lines, 4mm in from each edge. Plane the corners off, down to the gauge lines, and finish by rounding off the dowel until it fits into the holes in the bars.

MAKING THE WEDGES

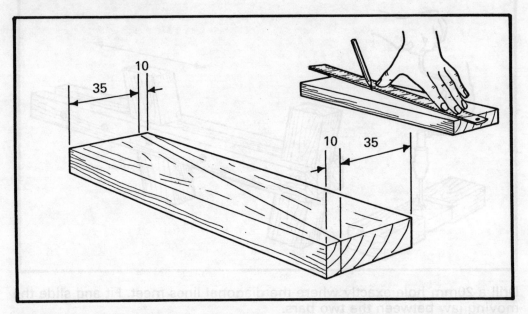

Take the timber prepared for the wedges. Square lines on the end grain of opposite corners, 10mm from each edge. Join these lines up on both sides with a ruler and pencil.

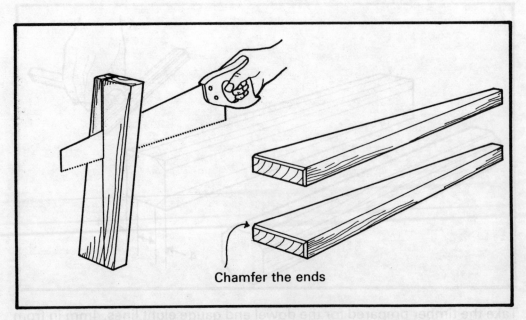

Chamfer the ends

Saw and plane the wedges to shape and chamfer the end grain. This will prevent them splitting when they are struck with a mallet.

USING THE CRAMPS

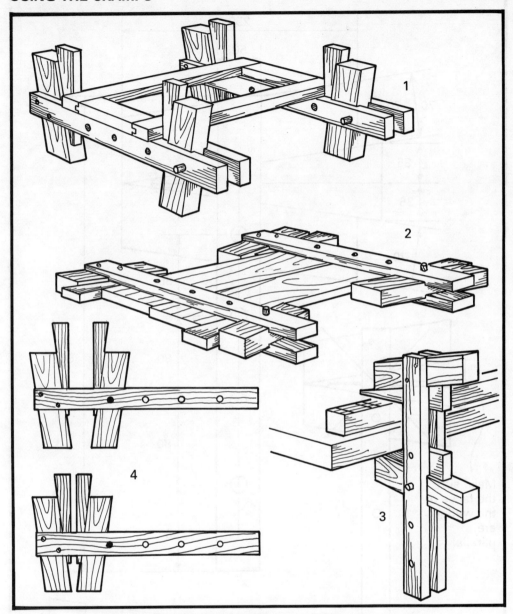

Tighten and loosen the cramps as shown on page 86.

Drawings 1 and 2:
Both the bench cramp and the sash cramp can be used in one of two ways for gluing up frames and table tops or panels.

Drawing 3:
The bench cramp can be used to hold timber to the workbench for mortising or sawing.

Drawing 4:
The moving jaw has two positions for each dowel hole.

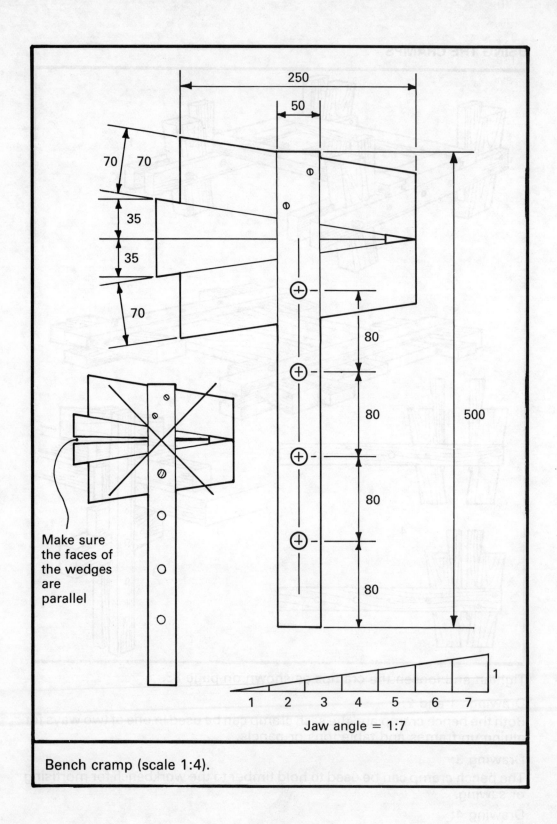

Bench cramp (scale 1:4).

Leg vice

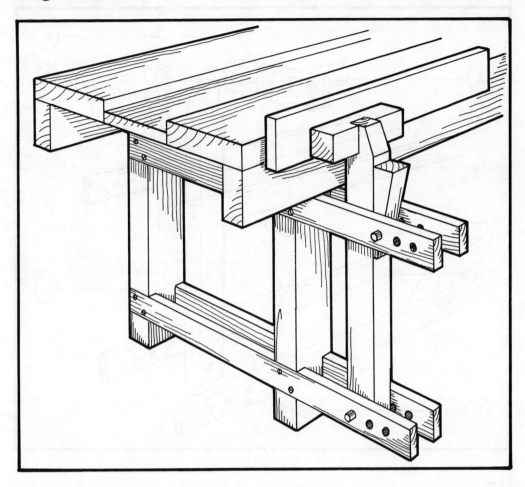

This is a very effective vice, made completely out of wood. Although it is quite a complicated tool to make, it will hold both large and small timbers, for planing, sawing and cutting joints.

The leg vice consists of two pairs of arms, fixed to both legs at one end of the bench. Each arm has three adjustment holes, drilled right through to take 25mm dowels. The jaw of the vice is attached to a leg which is pivoted on a dowel passing through one of three holes in the bottom pair of arms.

The vice is tightened by driving a wedge between the leg and the other dowel which fits into one of three holes in the top pair of arms.

In order to keep the jaw roughly parallel to the side of the bench when working with a thick piece of timber, the leg of the vice has to be adjusted. This is done by moving both the pivot dowel and the wedge-bearing dowel to one or other of the corresponding holes in both pairs of arms.

LIST OF PARTS

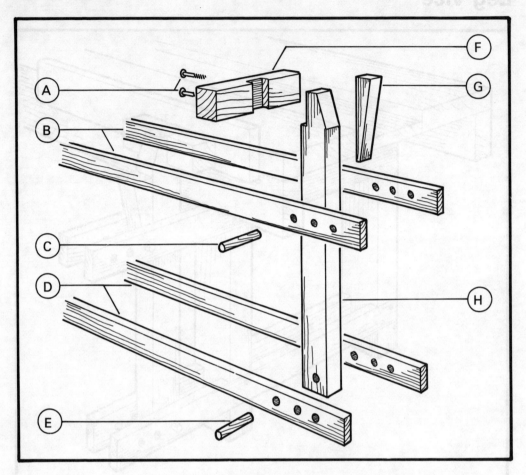

Part

A *Screws* are needed to fix the jaw to the leg.

B *Arms, top pair*. These have three adjustment holes drilled through to take the wedge bearing dowel. The arms are fixed securely to bench legs.

C *Wedge-bearing dowel*. This fits into one of three holes in the top pair of arms, and works against the wedge to put pressure on the leg.

D *Arms, bottom pair*. These have three adjustment holes drilled through to take the pivot dowel for the leg.

E *Pivot dowel*. This fits into one of the three holes in the bottom pair of arms and forms the pivot for the leg.

F *The jaw* is jointed and screwed to the leg of the vice.

G *The wedge* is driven between the leg and the wedge-bearing dowel, to put pressure on the workpiece.

H *The leg* is pivoted on one of three points on the bottom pair of arms.

CUTTING AND PARTS LIST

Part	Name	Quantity, Material and Dimensions (mm)
A	Screws	2 pcs. Wood screws 2″ (50mm) No.8
B & D	Arms	4 pcs. Timber 'Dim B' + 350 × 70 × 20
C & E	Dowels	1 pc. Timber 300 × 20 × 20
F	Jaw	1 pc. Timber 250 × 70 × 50
G	Wedge	1 pc. Timber 200 × 70 × 50
H	Leg	1 pc. Timber 750 × 70 × 50

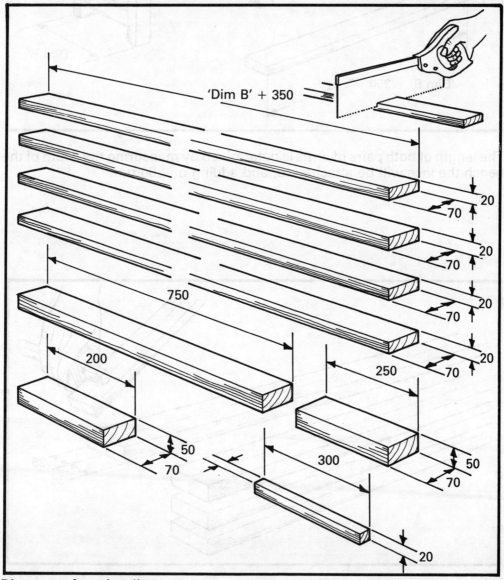

Diagram of cutting list.
'Dim B' refers to the width of the bench.

MAKING THE ARMS

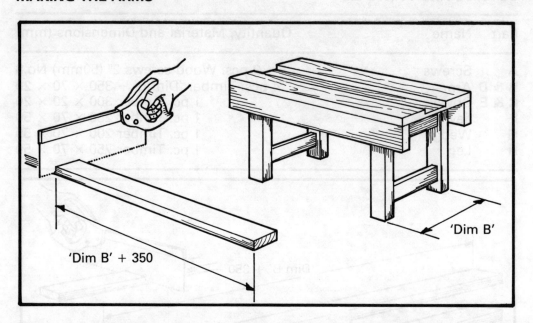

The length of both pairs of arms is determined by measuring the width of the bench the vice will be attached to, and adding on 350mm.

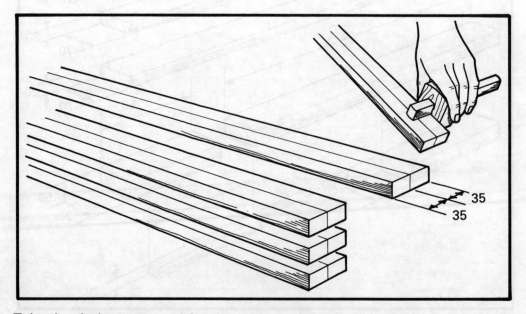

Take the timber prepared for the arms and gauge a centre line down both sides on all four pieces.

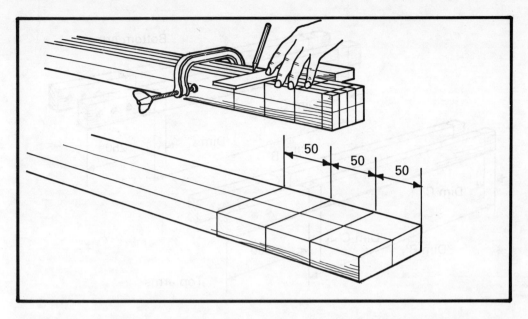

Cramp the arms together and square three lines across, 50mm apart, starting from one end. Remove the cramp and square the lines all round.

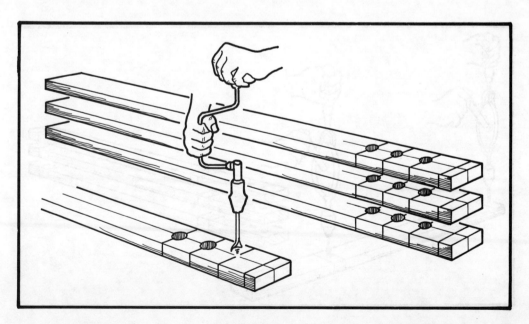

Use a carpenter's brace and a 25mm bit to drill three adjustment holes in all four arms.

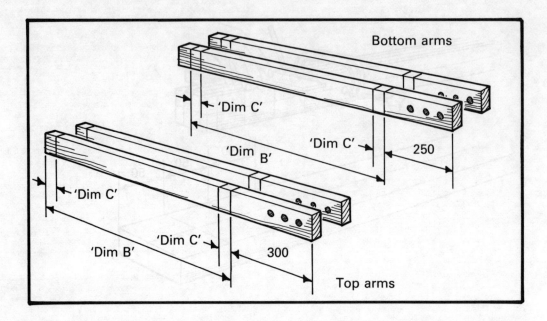

Use a pencil and a try square to mark lines all round the arms to the dimensions shown.

Note that the top arms project 50mm further from the bench than the bottom arms.

'Dim C' refers to the thickness of the bench legs.

Drill clearance holes into the arms and countersink for the heads of the screws to be used to fix the arms to the bench.

MAKING THE LEG

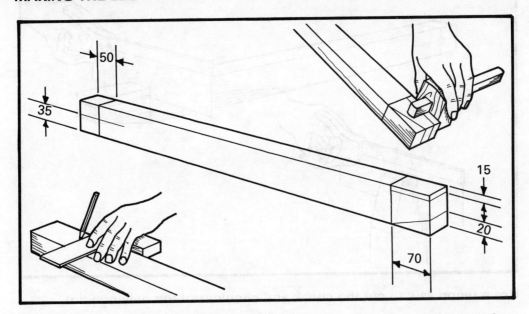

Take the timber prepared for the leg and square a line all round 50mm from one end. Gauge a centre line to mark the position of the pivot dowel.

Square another line all round 70mm from the other end, and gauge a line 15mm from one edge to mark the halving joint at the top of the leg.

Also mark a bevel on the other edge of the leg 20mm deep at the top.

Use a carpenter's brace and a 25mm bit to bore the pivot hole, and saw off the bevel at the top of the leg.

Use a tenon saw to cut the cheek and shoulder of the halving joint.

MAKING THE JAW

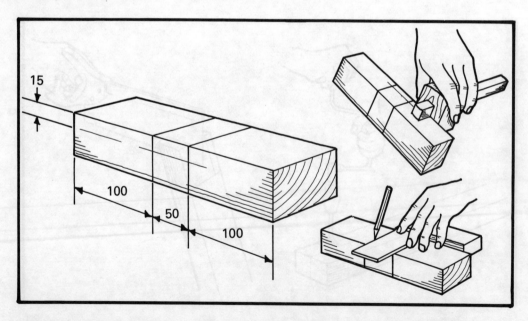

Take the timber prepared for the jaw, and square two lines 50mm apart in the centre of the block. Gauge the depth of the halving joint at 15mm.

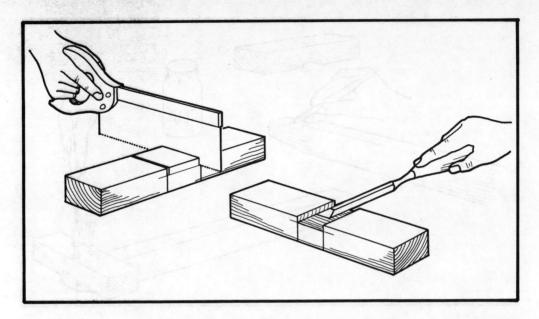

Saw down the sides of the joint with a tenon saw and remove the waste with a chisel.

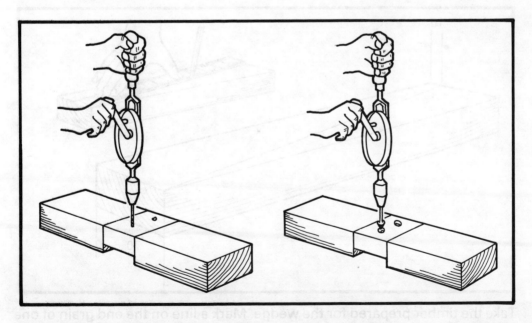

Use a wheel brace and bit to drill clearance holes through to the halving joint; countersink the holes well for the screw heads.

Clamp the leg and the jaw together and drill pilot holes for the screws.

Glue and screw the two parts tightly together.

MAKING THE WEDGE

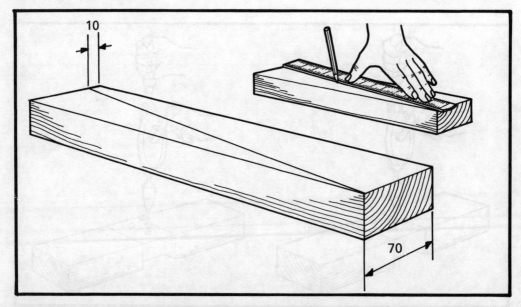

Take the timber prepared for the wedge. Mark a line on the end grain of one end, 10mm from the edge. On both sides join this point up to the opposite corner with a ruler and pencil.

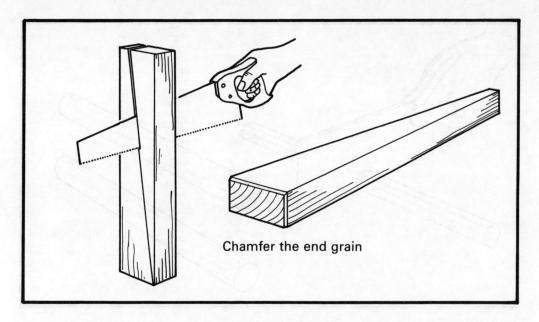

Chamfer the end grain

Saw and plane the wedge to shape. Chamfer the end grain, to prevent the wedge splitting when struck with a mallet.

MAKING THE DOWELS

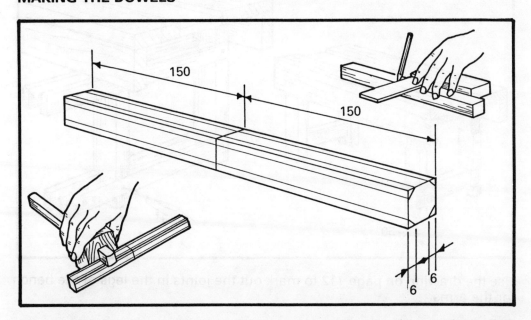

Take the timber prepared for the dowels and square a line all round exactly in the centre. Use a marking gauge to mark eight lines 6mm in from each edge.

109

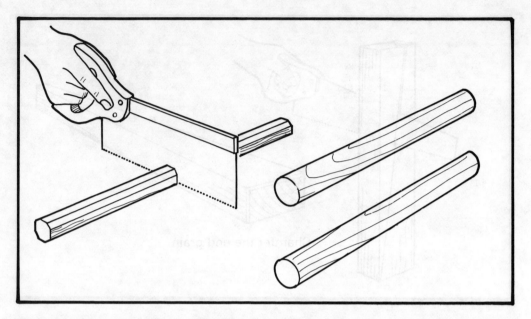

Plane the corners off down to the gauge lines, saw the dowel in half, plane them to a circular section and fit them into the adjustment holes in the arms.

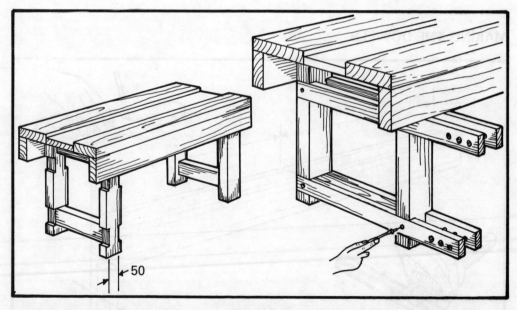

Use the drawing on page 112 to mark out the joints in the legs of the bench for the arms.

Cut the joints, and fix the arms to the bench with 2" (50mm) No.8 wood screws.

USING THE LEG VICE

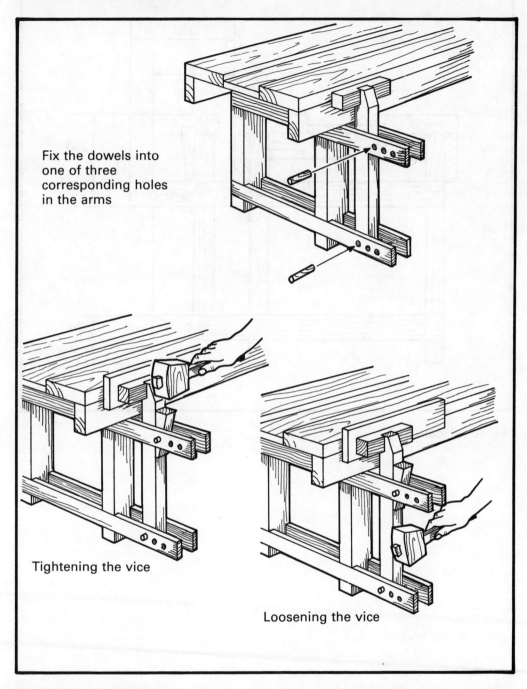

To tighten the vice, use a mallet to drive the wedge between the leg and the wedge-bearing dowel.

To loosen the vice, strike the bottom of the wedge with a mallet.

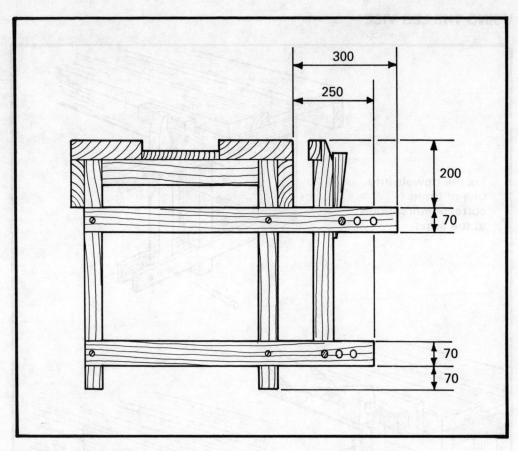

Leg vice (scale 1:4).

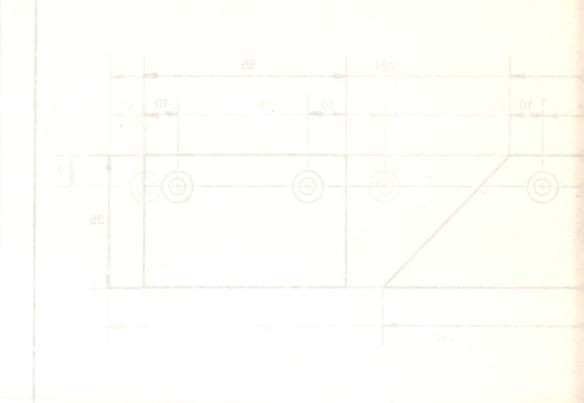

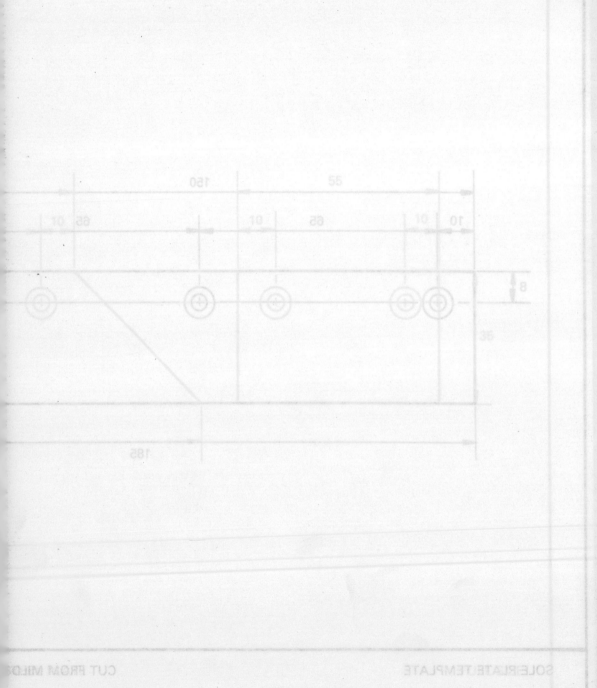